Handbook of Tritium NMR Spectroscopy and Applications

Handbook of Tritium NMR Spectroscopy and Applications

E. A. Evans* and *D. C. Warrell

Amersham International plc

J. A. Elvidge* and *J. R. Jones

Department of Chemistry
University of Surrey

A Wiley-Interscience Publication

JOHN WILEY & SONS

Chichester · New York · Brisbane · Toronto · Singapore

Library of Congress Cataloging in Publication Data
Main entry under title:

Handbook of tritium NMR spectroscopy and applications.

"A Wiley-Interscience publication."
Bibliography: p.
Includes indexes.
1. Nuclear magnetic resonance spectroscopy.
2. Tritium. 3. Radioactive tracers. I. Evans, E. Anthony (Eustace Anthony)
QD96.N8H36 1985 543′.0877 84–15273
ISBN 0–471–90583–6

***British Library Cataloguing in Publication Data*:**

Handbook of tritium NMR spectroscopy and applications.
1. Nuclear magnetic resonance spectroscopy
I. Evans, E. A.
538′.362 QC762

ISBN 0 471 90583 6

Typeset by Macmillan India Ltd, Bangalore.
Printed in Great Britain at The Pitman Press, Bath

Contents

Foreword

The need for a reliable routine analytical method of establishing the patterns of labelling in tritium labelled compounds became apparent in the early 1960s, when extensive uses of these tracer compounds developed in research in the life sciences. This need was subsequently satisfied by tritium nuclear magnetic resonance (^3H nmr) spectroscopy. For the detailed analysis of tritium labelled compounds, ^3H nmr is now arguably the method of choice, being non-destructive and rapid. Not only does the method reveal the positions and extent of tritium labelling, it also gives stereochemical information concerning the label. Tritium nmr is unquestionably a highly important analytical tool for monitoring the behaviour of tritium atoms in the wide ranging applications of tritium tracers.

Tritium nmr spectroscopy permits the ready determination not only of patterns of labelling but also of specific activity, purity and identity, and of other parameters, such as the stereochemical detail, vitally important in understanding the integrity and validity of tracer applications of tritium labelled compounds. In turn, patterns of labelling can be related to the synthetic or biosynthetic methods used for introducing tritium into organic molecules, to the behaviour of catalysts and to the detailed understanding of hydrogen–tritium exchange reactions.

This book reviews all aspects of the development and applications of ^3H nmr spectroscopy over 15 years of its routine use. It is written for all scientists using tritium labelled compounds as tracers, and for those contemplating their use, in order to provide them with a basis for better understanding the properties of such labelled compounds. Teachers and students should likewise find this text a useful guide to expanding their knowledge of the general principles of radiotracer techniques.

Following a brief introduction, highlighting the importance of ^3H nmr spectroscopy for tritium tracer studies, Chapter 1 deals with the theory of the method, the interpretation of spectra and other experimental aspects, emphasizing the importance of careful sample preparation and the special relationship of ^3H nmr spectral detail to the wealth of published data for proton spectra. Chapter 2 reviews the current methods for labelling compounds with tritium and the relationship of observed patterns of labelling to these methods. Chapter 3 describes applications of ^3H nmr spectroscopy to research in the life sciences which illustrate the power of the technique. Studies employing this analytical tool have revealed numerous interesting and indeed unexpected results in the behaviour of tritium atoms in labelled molecules. These studies have included applications of tritiated compounds in analytical and biochemical problems, in problems of catalysis and reaction mechanisms, and in other areas of scientific research.

The text is fully referenced, the many references to the literature including a bibliography of publications relating to ^{3}H nmr spectroscopy.

Overall the book is designed as a comprehensive handbook of information, references and basic text for spectroscopists, students and other users of radioisotopes in life sciences research, and reflects the knowledge accumulated over the 15 years of routine use of ^{3}H nmr spectroscopy in a collaborative programme between the Department of Chemistry at the University of Surrey, Guildford, and Amersham International plc.

John A. Elvidge
E. Anthony Evans
John R. Jones
David C. Warrell

Acknowledgements

Much of the work described in this text emerged from a collaborative project extending over more than 15 years. During this time, we have been grateful for the support of the following colleagues who have made valuable contributions in a variety of ways, so assisting us in developing the ^{3}H nmr spectroscopic technique and its applications.

From Amersham International:

Dr. John R. Catch (retired)
Dr. John C. Turner
Dr. Victor M. A. Chambers
Mrs. Hilary C. Sheppard
Dr. Ann-Marie Basketter (neé Davis)
Dr. Grahame L. Guilford
Dr. Brian S. Roughley

and from the University of Surrey:

Dr. J. M. A. Al-Rawi
Mr. J. P. Bloxsidge
Dr. D. Caddy
Mrs. L. Carroll
Mrs. M. Gower
Dr. D. K. Jaiswal
Dr. R. M. Lenk
Dr. R. B. Mane
Dr. C. O'Brien
Dr. K. D. Perring
Dr. M. S. Saieed
Dr. M. Saljoughian
Professor R. Thomas
Dr. Y. S. Tang
Dr. K. T. Walkin

For typing a draft and for setting the MS on a wordprocessor, we are especially grateful to Miss Prue Alder and Mrs Pat Stafford, respectively.

Introduction

Unquestionably, the relative ease with which even complex organic compounds can be labelled with tritium, at very high molar specific activity, has made tritium the most versatile beta emitting radionuclide for use in chemical and biological research. In many applications, tritium tracers fulfil a dual role, for tritium is not only a tracer for hydrogen but also for carbon structures. In fact for research in the life sciences, tritium is more often used as a tracer for carbon structures than in other ways. In order to interpret effectively, and so to understand, the behaviour of tritiated tracers under various environmental conditions, a knowledge of the precise distribution of tritium in labelled molecules is very necessary.

Unfortunately, because of the nature of carbon–hydrogen bonds, labelling methods for preparing tritium labelled compounds seldom lead to unambiguous labelling patterns [1]. Hitherto, the distribution of labelling in tritium labelled compounds has not been easily established, especially in complex molecules, because of non-specific isotope exchange reactions which can occur (unexpectedly) during preparation and during analysis by chemical and biochemical procedures. A non-destructive method of analysis became imperative, and so in 1968 a collaborative project was set up by Amersham International plc (formerly known as The Radiochemical Centre Limited, Amersham) with the Department of Chemistry at the University of Surrey, Guildford, in order to examine systematically tritium labelled compounds by tritium nuclear magnetic resonance spectroscopy. The initial instrumentation comprised a 60 MHz field-sweep nmr spectrometer adapted with a 64 MHz transmitter and receiver, and equipped with an elementary computer for limited spectral accumulation. The results were so useful and encouraging that in due course more elaborate instrumentation was warranted. This comprised a solid state 90 MHz Fourier transform instrument with a 96 MHz channel for ^{3}H observation, proton spin decoupling and a deuterium field-frequency lock with time-shared computer control and hard-disc storage of acquired data. With the advent of 'supercon' magnets and considerable advances in computer technology, the techniques of ^{3}H nmr have subsequently reached a high level of sophistication. The patterns of labelling in virtually any tritium labelled compound can now be readily determined at low isotopic abundance (e.g. 3×10^{-4} to 3×10^{-2} per cent. ^{3}H per site) by this direct, reliable, quick and non-destructive method.

Tritium (^{3}H or T) is fortunately an ideal nuclide for high resolution nuclear magnetic resonance studies. Its nucleus, the triton, like the proton, has a nuclear spin of $\frac{1}{2}$ and its high magnetogyric constant gives it a higher receptivity (sensitivity to detection) than any other nucleus, including the proton. However, in spite of these major advantages, there was initially a considerable psychological block to the development of ^{3}H nmr. The thought of spinning, at high speed, glass tubes containing multicurie quantities of tritiated compounds in very expensive nmr spectrometers seriously discouraged any rapid developments of the technique. The ^{3}H nmr signal from tritiated water was first observed in 1947 [1a], and in related measurements [9, 1b] several hundred curies were employed in specially shielded apparatus, but the true beginning of structural studies with ^{3}H nmr spectroscopy was in 1964 [2]. Shortly afterwards Tiers *et al.* [3] published the first 'high resolution' ^{3}H nmr spectrum of a tritiated organic compound, namely that of [*side-chain*-^{3}H]ethylbenzene. Although the spectrum indicated some potential for the method, the emphasis in the paper was on the contamination hazards which could arise from breakage of the nmr sample tube. This, coupled with the limitations of the instrumentation of the time, particularly as regards sensitivity, appeared to discourage further studies almost entirely.

During 1968 to 1971, the detailed practical evaluation (already referred to) revealed that tritium nmr spectroscopy could be made into a safe routine [4]. Most importantly, it became obvious that only millicurie (mCi) amounts of radioactivity were needed. The technique is now a most useful analytical tool, increasingly being applied to the study and utilization of tritium labelled compounds, as the present book relates.

Two consequences of the magnetic properties of the tritium nucleus make tritium nmr spectroscopy especially useful. Firstly, because of the high receptivity of the tritium nucleus, only reasonably small amounts of radioactivity (0.1 to 10 mCi per site, 3.7 to 370 MBq)* depending on available instrumentation are required to give a well defined spectrum. Secondly, because the chemical shifts of tritium nuclei are the same as those of hydrogen nuclei (protons) [5], the vast body of data available on proton chemical shifts can be applied directly to the interpretation of tritium spectra. No new correlations have to be determined, as may still often be the case in usage of ^{13}C nmr spectroscopy, for example.

As a rapid, direct and non-destructive method ^{3}H nmr has very real advantages to offer over other analytical approaches. Consequently ^{3}H nmr should always be considered in work involving tritium labelled compounds, provided that the available spectrometer sensitivity is sufficient. The information which tritium nmr gives so directly, on regiospecificity, quantitative distribution of the label and the stereochemistry of the label, cannot be obtained nearly so easily by any other method. Thus the use of millicuries rather than microcuries or lesser amounts of radioactivity may often be very justified in tritium tracer experiments, so that the new techniques may be applied.

A summary of the nuclear properties of some important isotopes is given in the Table 1. Table 2 lists essential information on tritium.

* See Appendix 1, page 219, for units of radioactivity.

Table 1. Nuclear properties of some isotopes used particularly in life sciences [6, 7]. (Reproduced by permission of The Royal Society of Chemistry.)

Isotope	Natural abundance (%)	Nuclear spin	Magnetic moment μ/μ_N	Magnetogyric ratio ($\gamma/10^7$ radians $T^{-1} s^{-1}$)	Resonance frequency (MHz at 2.114 T)	Relative sensitivity for equal numbers of nuclei at constant field	Half-life	Radiation	Maximum energy (MeV)
^{1}H	99.984	$\frac{1}{2}$	4.8371	26.7519	90.0	1.0	—	Stable	—
^{2}H (D)	0.0156	1	1.2125	4.1064	13.8	9.65×10^{-3}	—	Stable	—
^{3}H (T)	$< 10^{-16}$	$\frac{1}{2}$	5.1594	28.5336	96.0	1.21	12.43 y*	β^-	0.018
^{12}C	98.89	0	—	—	—	—	—	Stable	—
^{13}C	1.11	$\frac{1}{2}$	1.2162	6.7263	22.6	1.59×10^{-2}	—	Stable	—
^{14}C	$< 10^{-10}$	0	—	—	—	—	5,730 y	β^-	0.156
^{14}N	99.63	1	0.5706	1.9324	6.50	1.01×10^{-3}	—	Stable	—
^{15}N	0.37	$\frac{1}{2}$	−0.4902	−2.7107	9.12	1.04×10^{-3}	—	Stable	—
^{16}O	99.76	0	—	—	—	—	—	Stable	—
^{17}O	0.037	$\frac{5}{2}$	−2.2398	−3.6266	12.20	2.91×10^{-2}	—	Stable	—
^{18}O	0.20	0	—	—	—	—	—	Stable	—
^{31}P	1.00	$\frac{1}{2}$	1.9581	10.8290	36.43	6.63×10^{-2}	—	Stable	—
^{32}P	0	1	−0.3568	−1.2084	4.06	2.46×10^{-4}	14.3 d	β^-	1.710

* Currently reassessed value is 4,540 days [8].

Table 2. Some properties of tritium (^{3}H) and related information

Mass	3
Radiation emitted	Beta (100%)
Half-life	12.43 y (4,540 d)
Energy (maximum)	0.0186 MeV
Range of particles:	
in air	4.5–6 mm
in water	6 nm
Production method	^{6}Li (n, α) ^{3}H
Decay product (stable)	[^{3}He]helium
Available isotopic abundance	100%
Maximum specific activity	28.76 Ci mmol^{-1}
Volume of 1 Ci of $^{3}H_2$ gas at STP	0.385 ml
Dissociation energy for $^{3}H_2 \rightarrow 2\,^{3}H$	4.59 eV
Common specific activities for compounds	1–100 Ci mmol^{-1}
Method of measurement	Liquid scintillation
(Efficiency)	(40%)

CHAPTER 1

Experimental aspects of ^{3}H nmr spectroscopy

1. Nuclear magnetic properties of tritium

The triton, the nucleus of tritium, has a spin quantum number $I = \frac{1}{2}$, so that it is a suitable nucleus for high resolution magnetic resonance study, like the proton, as explained below. The $(2I + 1)$ quantum rule indicates two energy states or orientations for the nuclear magnetic vector in an applied field and therefore one transition, so that, with correct spectrometer operating conditions, the integrated spectral line intensities give a direct quantitative measure of the amounts of tritium responsible. The spin quantum number $I = \frac{1}{2}$ also implies that the nucleus has spherical symmetry so the *only* property dependent upon orientation is the magnetic moment. Hence the relaxation of tritons from their upper spin state will be induced only by *magnetic* fields. These need to have components perpendicular to the applied magnetic field (of the spectrometer) and to be oscillating with the nuclear precessional or Larmor frequency. Such fields will arise in the sample by chance from the collection of nuclear spins, but this mechanism of spin–lattice relaxation will be slow because of the stringent requirements. The resultant relatively long upper-state lifetime of the triton will lead, by the uncertainty principle, to a very narrow linewidth, so that high resolution nmr spectra are obtained. There is a close analogy with the proton, which has a rather similar (though lower) resonance frequency and analogous stringent requirements for relaxation. It follows that nmr spectrometer operating conditions such as power level, pulse length, etc., which are optimum for ^{3}H nmr spectroscopy, will in general be similar to those for ^{1}H nmr spectroscopy.

Another property of the tritium nucleus which facilitates ^{3}H nmr is the high nuclear magnetic moment μ_T which causes the magnetogyric constant $\gamma\!\!\!-_T$* $(= \mu_T/Ih = 4.5414 \times 10^7$ Hz T^{-1}) to be higher than for any other nucleus including the proton ($\gamma\!\!\!-_H = 4.2577 \times 10^7$ Hz T^{-1}). Thus at 2.114 T, at which field the ^{1}H nmr frequency is 90 MHz, the ^{3}H nmr frequency will be $4.5414 \times 2.114 \times 10^7 = 96$ MHz. This higher nmr frequency results in slightly better spectral dispersion (by the factor $\gamma_T/\gamma_H = 1.06664$) in ^{3}H nmr spectra compared with ^{1}H

* $\gamma\!\!\!- = \gamma/2\pi$, a symbol change consequent upon use of hertz instead of radians per second.

spectra at the same field. Also increased by the same factor are the coupling constants for tritons to other nuclei. A further consequence of the uniquely high magnetogyric constant for the triton is that it gives this nucleus the highest receptivity or sensitivity to nmr detection. This is expressed as $(\gamma_T/\gamma_H)^3 = 1.21$ as compared to 1.00 for an equal number of protons at the same field. The high receptivity is clearly an advantage for the nmr study of tritium in tritiated compounds where the isotope will frequently be at rather low abundance, such as 0.1 to 0.00035 per cent. per site, corresponding to 30 mCi to 100 μCi of radioactivity per site (1.11 GBq to 3.7 MBq).

Briefly, then, the nucleus of tritium is ideal for high resolution nmr examination. Being an isotope of hydrogen, tritium has virtually the same chemical shifts, so no new reference data are needed. Its receptivity is the highest of any magnetic nucleus; coupling constants will be slightly larger than the corresponding proton coupling constants but predictable; under appropriate conditions the integrated line intensities will give a quantitative tritium analysis; and nmr spectrometer operating conditions will not be significantly different from those for ^{1}H nmr except for the higher resonance frequency. All these features serve to make ^{3}H nmr spectroscopy an especially attractive technique for determination of structures and sites of labelling of tritiated compounds, for rapid determination of the relative amounts of tritium at different sites and for investigation of the stereochemistry of tritium labelling.

2. Chemical shifts

Chemical shifts, which describe the positions of the resonance lines in nmr spectra, are obviously important for their correlation with structure (see Appendix 2, page 220, and Figure 1). Because the frequency of a resonance line is a function of the spectrometer, namely its magnetic field strength, chemical shifts are normally given as δ values in parts per million (p.p.m.) from a stated reference so that they are independent of the instrumentation. Thus the chemical shift δ_X of a nucleus X in a sample s is related to the frequency of the observed nuclear resonance line ν_X^s and to that of the relevant internal reference ν_X^r as follows:

$$\delta_X = \frac{\nu_X^s - \nu_X^r}{\nu_X^r} \quad \text{p.p.m.} \tag{1}$$

where the separation $(\nu_X^s - \nu_X^r)$ is measured in hertz and ν_X^r is measured in megahertz (and is generally taken to be the nominal operating frequency of the spectrometer).

The condition for nuclear magnetic resonance of any nucleus X at constant field B_0 is given by equation (2), where σ_X is the local screening or shielding constant:

$$\nu_X = \gamma_X B_0 (1 - \sigma_X) \tag{2}$$

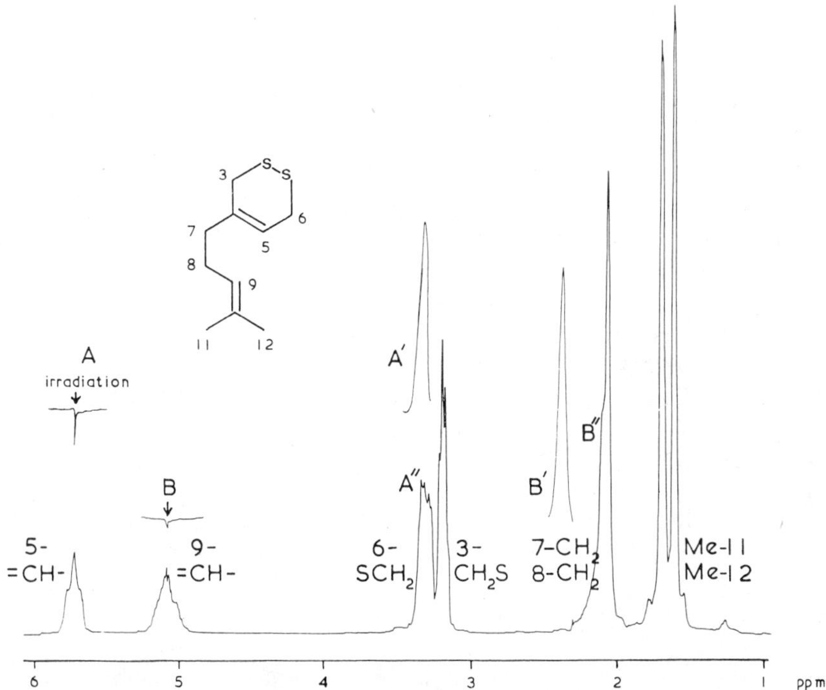

Figure 1. Typical nmr spectrum (^{1}H) showing correlation of signals with structural features, as deduced from tables (see Appendix 2) and in part from effects A′ and B′ on signals A″ and B″, respectively, produced by specific decoupling applied at A and at B

Combination of equation (2) with equation (1) in the form $\delta_X = (\nu_X^s/\nu_X^r) - 1$, for both hydrogen and tritium, gives

$$\frac{\delta_T}{\delta_H} = \frac{[(1-\sigma_T^s)/(1-\sigma_T^r)]-1}{[(1-\sigma_H^s)/(1-\sigma_H^r)]-1} \tag{3}$$

Because the shielding of a hydrogen nucleus in a compound in solution is almost entirely a function of the local molecular environment, there is virtually no change in screening at an isotopically replaced hydrogen nucleus, so that

$$\sigma_T^r \simeq \sigma_H^r \quad \text{and} \quad \sigma_T^s \simeq \sigma_H^s$$

As a consequence, equation (3) becomes the near equality

$$\frac{\delta_T}{\delta_H} \simeq 1 \tag{4}$$

Much earlier, Duffy had concluded that the ratio of the hydrogen isotopic nuclear screenings must be very close to unity [9]. Experimental results made with a simple field-sweep spectrometer (Table 3) confirmed that this is indeed the case when chemical shifts for ^{1}H and ^{3}H at the same site (in partially tritiated

Table 3. ^3H and ^1H chemical shifts (p.p.m.) from internal tritiated water, used as solvent for the stated partially monotritiated compounds (negative shifts are upfield). (Reproduced by permission of The Royal Society of Chemistry.)

Compound (position tritiated)	δ_T	δ_H
Sodium acetate (2)	−2.91	−2.87
Acetic acid (2)	−2.77	−2.72
Acetone (1)	−2.66	−2.64
2-Picoline (1′)	−2.58	−2.59
Acetonitrile (2)	−2.25	−2.22
Propionitrile (2)	−2.21	−2.22
Dimethyl sulphoxide (1)	−2.08	−2.09
Sodium malonate (2)	−1.60	−1.63
Nitromethane (1)	−1.17	−1.18
Malononitrile (2)	−0.75	−0.76
Diethyl malonate (2)	−0.50	−0.47
2-Methylresorcinol (4)	1.47	1.50
Imidazole (2)	2.63	2.62
Benzimidazole (2)	2.94	2.91
Purine (8)	3.09	3.07
Pyridine-1-oxide (2)	3.42	3.42
Benzoxazole (2)	3.63	3.63
Benzothiazole (2)	4.40	4.37
Chloroform (1)	5.07	5.16
Benzoselenazole (2)	5.28	5.34

compounds) are measured under the same conditions [5]. Similar observations have been reported for other isotopic pairs, e.g. ^1H and ^2H [10] and ^{14}N and ^{15}N [11, 12]. The importance of this finding lies in the fact that the vast compilation of ^1H chemical shift data in the literature will also apply to the prediction and assignment of ^3H nmr spectra, so that there is little or no interpretive hindrance to applications of ^3H nmr spectroscopy.

The chemical shifts of the magnetic nuclei in a substrate, measured from a recognized internal reference, are of course somewhat dependent upon the nature of the solvent, the temperature and the concentration of the solution. Hence the operating conditions should always be stated. It is best to use solvents which are more or less magnetically isotropic (e.g. $CDCl_3$, deuterated dioxan, cyclohexane or ideally CCl_4) and to keep the concentration of solute low, i.e. 5 per cent. or less, in order to facilitate comparative measurements and to obtain reproducible results. For stronger solutions and for solutions in water or in highly anisotropic solvents such as benzene or pyridine, the concentration must be stated or the chemical shifts will be of little value. Deuterated solvents are frequently used because field-frequency locking to a solvent ^2H resonance is a common feature of FT nmr spectrometers, and in any case such solvents allow comparison of ^1H spectra with ^3H spectra on the same sample. For ^3H-labelled compounds of high specific activity, the concentration will almost inevitably be very low and so the chemical shifts, even in the less isotropic solvents such as dimethyl sulphoxide, acetone and methanol, will be easily reproducible, as in isotropic solvents (i.e. δ values will not vary significantly over reasonable ranges of low concentrations of substrate and will tend to shifts at zero concentration). If there is carrier present, the concentration of substrate might be high and so should be stated, and comparisons made only at equal molar concentrations. With the highly anisotropic solvents, benzene or pyridine, the chemical shifts of solutes are likely to be strongly dependent on concentration, even at high dilution. Thus for [2-^{3}H]mevalonolactone (discussed in another connection later—see Figure 5) in d_6-benzene, the chemical shifts of the non-equivalent labelled positions vary in the concentration range 0.02 to 2.0 per cent. (w/v) as follows:

Concentration (%)	δ (p.p.m.)
0.02	1.837, 2.113
0.20	1.850, 2.156
2.00	1.922, 2.331

Solvents such as benzene are effectively 'shift' solvents and it is most important, for use of derived data, to know the concentration. Equally, data for neat liquids must not be compared with those for dilute solutions because of the relatively large differential chemical shifts, which may lead to line crossover, i.e. the order of close chemical shifts may change with dilution (see pages 185 to 187), as is often encountered in use of lanthanide shift reagents. Such matters are, of course, well appreciated by nmr spectroscopists.

3. The Larmor frequency ratio

The experimental finding that hydrogen isotopic chemical shifts are the same, i.e. $\delta_T/\delta_H \simeq 1$, equally means (equation 1) that the ratio of the resonance or Larmor frequencies, ω_T/ω_H, for any site at constant field, should be virtually constant. Such frequency measurements can readily be made with high precision on a Fourier transform, pulse nmr spectrometer. Using a wide range of partially monotritiated compounds, measurements of nmr line frequencies (Table 4) gave a mean value of 1.06663975 ($\pm 3 \times 10^{-8}$) for the ratio [5].

In subsequent work [13, 14] small discrepancies (generally < 0.1 p.p.m.) between ^{3}H and ^{1}H chemical shifts were found for [G-^{3}H]alkylbenzenes. Thus ^{3}H alkyl resonances were 0.01–0.03 p.p.m. to high field of corresponding ^{1}H chemical shifts measured on the same sample, whilst the reverse frequently held for the aryl resonances. To make sure that these differences could not be attributed to experimental error, the Larmor frequency ratio was measured [14] as precisely as possible for a further series of selected tritiated compounds (Table 5). The values clearly fell into three groups, depending upon the carbon–hydrogen

Table 4. Measurements of the ratio of ^3H and ^1H Larmor frequencies.

Compound (position tritiated)	ω_T/ω_H
Sodium acetate (2)	1.066639716
Acetone (1)	1.066639744
Acetonitrile (2)	1.066639734
Dimethyl sulphoxide (1)	1.066639758
Nitromethane (1)	1.066639748
Benzimidazole (2)	1.066639779
Benzoxazole (2)	1.066639731
Benzothiazole (2)	1.066639781
Chloroform (1)	1.066639718
Acetophenone (Me)	1.066639747
Water	1.066639765

Table 5. Dependence of the Larmor frequency ratio on bond hydridization. (Reproduced by permission of John Wiley & Sons Ltd.)

Compound (position tritiated)	ω_T/ω_H	C–H bond type
Phenylacetylene (1)	1.066639738	sp^1
Methylbutynol (4)	1.066639740	sp^1
Undec-10-ynoic acid (11)	1.066639732	sp^1
Benzene (G)	1.066639783	sp^2
Collidine (3, 5)	1.066639784	sp^2
Cinnamic acid (α)	1.066639780	sp^2
Isobutyric acid (2)	1.066639681	sp^3
Isobutanol (2)	1.066639694	sp^3
Pentan-3-one (2, 4)	1.066639718	sp^3
Toluene (Me)	1.066639719	sp^3
Collidine (2, 6-Me)	1.066639724	sp^3
Collidine (4-Me)	1.066639722	sp^3

bond hybridization. Evidently, the ratio of the isotopic nuclear screenings is not constant but varies detectably with bond type, as indeed has been noted for deuterated compounds [15, 16].

The slight differential screening of ^3H and ^1H nuclei leads to problems in the choice of a best possible reference for ^3H nmr spectroscopy. Any given choice would lead to its own characteristic set of isotope effects [14]. However, because of the wide use being made of tetramethylsilane as a reference in ^1H nmr spectroscopy, the corresponding partially monotritiated compound appeared highly appropriate. It was prepared via the following sequence of reactions

$$Me_3SiCH_2Cl \xrightarrow{Li} Me_3SiCH_2Li \xrightarrow{MeOT/THO} Me_3SiCH_2T$$

and the Larmor frequency ratio determined (Table 6). Interestingly and fortunately the average value found ($1.066639739 \pm 2 \times 10^{-9}$) is very close to the mean value ($1.06663975 \pm 3 \times 10^{-8}$) previously derived from an arbitrary range of tritiated compounds, so our earlier measurements of ^{3}H chemical shifts do not need correction.

Table 6. Measurements of the ratio of ^{3}H and ^{1}H Larmor frequencies for partially monotritiated tetramethylsilane at 25 °C. (Reproduced by permission of John Wiley & Sons Ltd.)

Solvent	ω_T/ω_H
d_6-Benzene	1.066639737
	1.066639740
	1.066639739
d_1-Chloroform	1.066639740
Average	$1.066639739 \pm 2 \times 10^{-9}$

Having obtained a sufficiently precise value for the ratio of the Larmor frequencies of the two nuclei, ^{1}H and ^{3}H, in partially monotritiated tetramethylsilane, it is now possible to reference any ^{3}H nmr spectrum accurately and conveniently without the need for an actual tritium reference being present at all. Indeed the routine use of [^{3}H]TMS or other ^{3}H-reference would be highly undesirable. Instead, ordinary tetramethylsilane (TMS) or sodium 4,4-dimethyl-4-silapentane-1-sulphonate (DSS) is added to the sample. This is current practice in ^{1}H nmr spectroscopy, providing the generally accepted internal proton reference and of course in ^{3}H nmr spectroscopy facilitating the taking of the ^{1}H nmr spectrum, when required. The accurate resonance frequency of the internal proton reference signal, observed with the ^{1}H radiofrequency channel and available from the spectrometer output, is then multiplied by the Larmor frequency ratio 1.06663974 to provide the reference frequency for the ^{3}H nmr spectrum. This corresponds to a ^{3}H nmr signal which would have arisen if monotritiated TMS or DSS had been present in the sample. The appropriate commands to the computer then place this ghost internal reference signal upon the right-hand ordinate of the chart paper. The chemical shifts of the ^{3}H nmr signals from the sample are then obtained direct from the printout or the plotted spectrum in the usual way (see Figure 2).

Modern nmr spectrometers derive all the required frequencies from a master clock, so these frequencies are extremely stable with respect to one another. Provided the field is locked to a given hetero resonance from the sample (e.g. ^{2}H) and the sample is chemically stable during observation, successive ^{3}H and ^{1}H measurements will be strictly comparable, i.e. made at exactly the same field. Initially, the relationship between the frequencies used in the observation

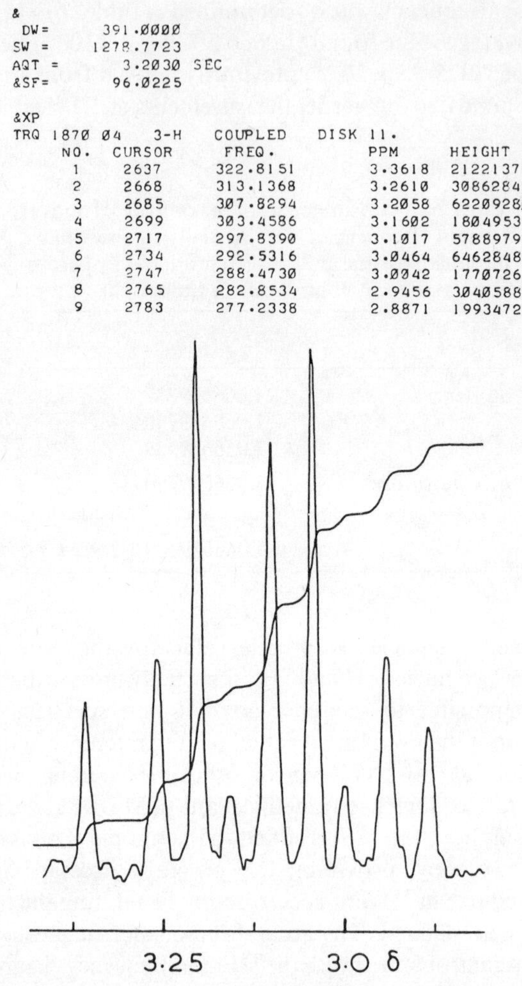

&
DW = 391.ØØØØ
SW = 1278.7723
AQT = 3.2Ø3Ø SEC
&SF= 96.Ø225

&XP
TRQ 187Ø Ø4 3-H COUPLED DISK 11.

NO.	CURSOR	FREQ.	PPM	HEIGHT
1	2637	322.8151	3.3618	21221376
2	2668	313.1368	3.261Ø	3Ø862848
3	2685	3Ø7.8294	3.2Ø58	622Ø928Ø
4	2699	3Ø3.4586	3.16Ø2	18Ø49536
5	2717	297.839Ø	3.1Ø17	57889792
6	2734	292.5316	3.Ø464	6462848Ø
7	2747	288.473Ø	3.ØØ42	177Ø7264
8	2765	282.8534	2.9456	3Ø4Ø5888
9	2783	277.2338	2.8871	1993472Ø

Figure 2. Example of computer printout accompanying a ^{3}H nmr spectrum. From top left,

DW = dwell time/μs
where DW/s = 1/(2 × spectral width/Hz)
and (2 × spectral width/Hz) = sampling frequency/Hz

SW = spectral width/Hz

AQT = acquisition time/s
= no. of computer addresses × DW/s
= (4,096 + 4,096*) × DW/s
where 4,096* zero-filled addresses are added to the 4,096 addresses used for acquisition before Fourier transformation. The procedure improves definition though not, of course, resolution.

&SF = spectrometer frequency/MHz
= frequency of ghost reference/MHz, derived from [^{1}H]TMS as $\nu_{TMS} \times (\omega_T/\omega_H)$

channels needs to be determined carefully, as well as the relationship between these and the offset frequencies used to set the pulse frequency. Fortunately, frequency comparison is straightforward and precise. Knowing the ^3H pulse frequency and its computer address, the calculated ^3H reference frequency (or ghost frequency) can be assigned to its correct address and so be treated as the internal reference point for the spectral display.

4. Coupling constants

The magnitudes of tritium–tritium and proton–tritium spin coupling constants can provide useful stereochemical information. Theory shows that for first row elements in the periodic table the spin–spin coupling constants are proportional to the product of the magnetogyric ratios of the coupled nuclei [17]. Thus the proton and tritium isotopic spin–spin coupling constants are related as follows:

$$J_{TT} = J_{HT}(\gamma_T/\gamma_H) = J_{HH}(\gamma_T/\gamma_H)^2 \tag{5}$$

(or)

$$J_{HT} = J_{HH}(\gamma_T/\gamma_H) \tag{6}$$

where $(\gamma_T/\gamma_H) \approx 1.0666$

Exactly analogous expressions connect the values of hydrogen isotopic coupling constants to other nuclei; e.g.

$$J_{CT} = J_{CH}(\gamma_T/\gamma_H) \tag{7}$$

The magnetogyric ratio γ refers to the bare nucleus and as the ratio cannot be measured directly it is replaced by the experimental Larmor frequency ω, which inevitably incorporates the local screening effects and hence isotope effects. This is well illustrated by comparison of some of the earliest measured carbon–tritium coupling constants with the values calculated from the corresponding carbon–proton coupling constants (Table 7). The isotope effects, though rather small, are measurable [18], being just outside the experimental error of measuring chemical shifts.

For the determination of geminal proton–proton coupling constants where the protons are equivalent (isochronous), the conventional approach has been to

---

&XP = instruction to computer to commence printout of spectral data

Above the columns, the sample number, type of spectrum and magnetic disc store are specified. The spectral lines are numbered, here 1–9, and given alongside are their frequencies from the reference and the corresponding chemical shifts, δ p.p.m., which are the quotients of the line frequencies and the reference frequency. The heights are the actual numbers stored in the addresses of the peak tops. The steps in the integral trace across the spectrum are driven from those numbers plus the numbers in the two or three adjacent addresses on each side of a peak address, so that the steps are a measure of the integrated peak intensities. (Reproduced by permission of John Wiley & Sons Ltd.)

Table 7. Direct carbon–proton and carbon–tritium coupling constants measured from the ^1H and ^3H nmr spectra, and primary isotope effects. (Reproduced by permission of The Royal Society of Chemistry.)

Compound (position tritiated)	J_{CH} (Hz) (measured)	J_{CT} (Hz) (measured)	J_{CT} (Hz)* (calculated)	Primary isotope effect (Hz)
Acetonitrile (2)	136.2	145.02	145.28	−0.26
Acetophenone (Me)	127.34	136.09	135.83	+0.26
Chloroform (1)	209.12	221.51	223.06	−1.55
Diethyl malonate (2)	132.28	140.90	141.09	−0.19
Sodium acetate (2)	127.12	135.78	135.59	+0.19

* $J_{CT} = J_{CH} \times 1.06664$.

examine the corresponding monodeuterated compounds. Measurement of $^2J_{HD}$ in the ^1H nmr spectrum and application of the following equation

$$J_{HH} = J_{HD}(\gamma_H/\gamma_D) \tag{8}$$

where $(\gamma_H/\gamma_D) \approx 6.5144$

then provides a value for $^2J_{HH}$. Here the measured quantity is small and the multiplying Larmor ratio is relatively large, so that the measurement errors are magnified. A superior approach is to prepare the partially monotritiated compound (see Table 8) and measure $^2J_{TH}$ in the ^3H nmr spectrum. The calculation of the required $^2J_{HH}$ (equation 6) is then achieved with improved precision over the previous method because the observed quantity is larger and the multiplying factor is near unity. Indeed the resultant precision will be about the same as for a direct measurement of J_{HH} where this is possible.

Table 8. Geminal tritium–proton coupling constants measured from ^3H nmr spectra, and calculated proton–proton values. (Reproduced by permission of The Royal Society of Chemistry.)

Compound (position tritiated)	J_{HT} (Hz) (observed)	J_{HH} (Hz) (calculated)*
Acetone (1)	15.27	14.32
Acetonitrile (2)	17.15	16.08
Dimethyl sulphoxide (1)	13.24	12.41
Diethyl malonate (2)	16.84	15.79
Nitromethane (1)	13.69	12.83
Sodium acetate (2)	14.95	14.02

* $J_{HH} = J_{HT}/1.06664$.

When geminal hydrogens are non-isochronous (have different chemical shifts), e.g. when they are prochiral, the geminal coupling constant $^2J_{HH}$ can of course be measured. Thus for benzyl methyl sulphoxide partially tritiated in the methylene group, both $^2J_{HH}$ and $^2J_{HT}$ could be measured and so compared direct [14]. The measured value for $^2J_{HH}$ was 0.11 Hz greater than that calculated from $^2J_{HT}$ measured under the same conditions. The discrepancy, which was the tritium primary isotope effect on this particular geminal coupling, was thus obtained with fair precision. Such isotope effects are of theoretical interest; occasionally they can assist in interpretation of the ^{3}H nmr spectra of multitritiated species.

5. Isotope effects on chemical shifts

The increasing need for tritiated compounds of very high specific activity (three or more ^{3}H atoms per molecule) has led to the development of new synthetic routes. Methylation employing tritiated methylating reagents is now widely used because such reagents are readily derived from tritiated methanol. This in turn is available as CT_3OH as a result of the direct catalytic reduction of carbon monoxide and/or carbon dioxide with tritium gas. Methylation of, for example, desmethylflunitrazepam with high specific activity methyl iodide gave [*N-methyl*-^{3}H]flunitrazepam (**I**). The ^{3}H nmr spectrum (Figure 3a) taken with ^{1}H broad-

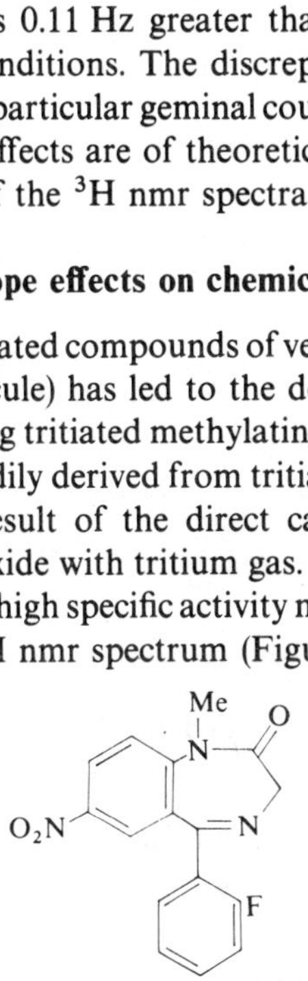

Me O N O$_2$N N F

(**I**)

band decoupling showed a strong CT_3 singlet signal at $\delta = 3.36$. The ^{1}H-coupled ^{3}H nmr spectrum (Figure 3b) showed signs of a weak doublet at the base of the singlet and the resolution-enhanced spectrum (Figure 3c) confirmed this by showing a resolved, weak, doublet signal centred at $\delta = 3.39$ (with $J_{HT} = 13.4$ Hz), surrounding the strong singlet at $\delta = 3.36$. The relative proportions of the compound containing the CHT_2 and CT_3 groups were 6 and 94 per cent., respectively. Although the resolution enhancement procedure necessarily distorts the line shape and so may affect intensity, no appreciable relative effects are to be expected when signals from one methyl group are being compared [19].

A different example of a highly tritiated compound is [*methyl*-^{3}H]thymine (**II**),

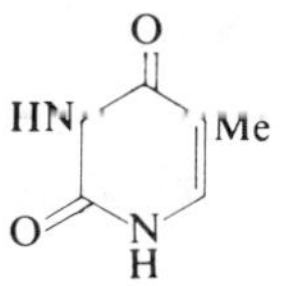

O HN Me O N H

(**II**)

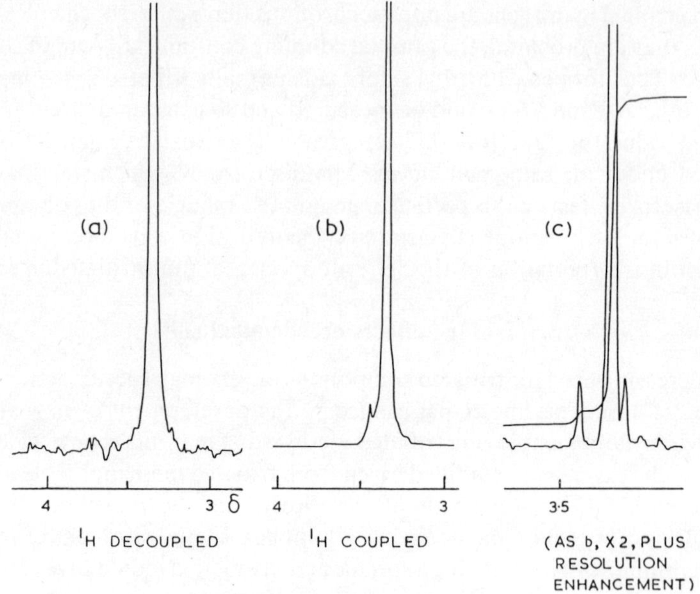

Figure 3. ^{3}H nmr spectra of [*N-methyl*-^{3}H]flunitrazepam in d_6-DMSO. (Reproduced by permission of John Wiley & Sons Ltd.)

prepared by catalysed reduction of 5-formyluracil with tritium gas. The ^{3}H nmr (^{1}H-decoupled) spectrum (Figure 4a) showed three partially merged lines in the $\delta = 1.8$ region. Resolution enhancement separated the lines (Figure 4b) and so allowed their relative intensities at $\delta = 1.79, 1.81$ and 1.84 to be measured as 29, 54 and 17 per cent., respectively. The origin of these lines from CT_3, CHT_2 and CH_2T methyl groups, respectively, was demonstrated by the ^{1}H-coupled resolution enhanced ^{3}H nmr spectrum (Figure 4c) which clearly shows a singlet (CT_3), doublet (CHT_2) and triplet (CH_2T) pattern. The lines in the spectrum (Figure 4c) were somewhat broad; further enhancement (at the cost of signal-to-noise) revealed the long-range coupling to the 6-proton, $^4J_{TH} = 1.1$ Hz, as shown (inset d) for the CT_3 signal [19].

The ^{3}H nmr spectra of such highly tritiated species, taken with the corresponding ^{1}H nmr spectra, allow the determination of the specific radioactivity of the product. This is a potentially useful alternative method to serial dilution and counting, especially for those compounds which do not possess an appropriate absorption spectrum or other properties which enable very low chemical concentrations to be measured with the required accuracy.

6. ^{3}H NMR signal intensities; relaxation and nuclear Overhauser effects

As already explained (page 1), the integrated intensities of ^{3}H nmr signals are proportional to the numbers of nuclei responsible for those signals. Hence

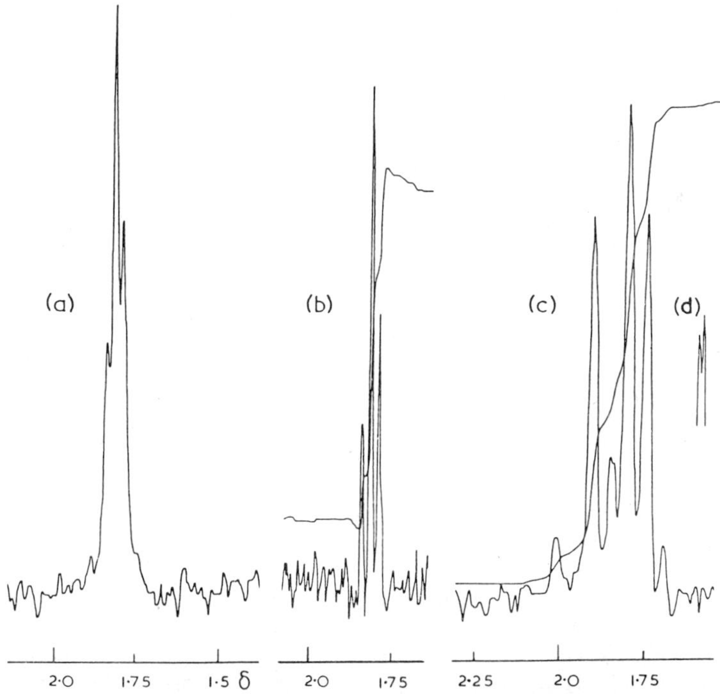

Figure 4. ^3H nmr spectra of [*methyl*-^3H]thymine in D_2O: (a) ^1H decoupled; (b) as (a) with resolution enhancement; (c) ^1H coupled, resolution enhanced; (d) further enhancement of CT_3 singlet in (c). (Reproduced by permission of John Wiley & Sons Ltd.; (c), (d) reproduced from W. P. Duncan and A. B. Susan, *Synthesis and Applications of Isotopically Labeled Compounds* by permission of Elsevier Science Publishers B. V.)

a ^3H nmr spectrum gives both quantitative information as well as qualitative structural information (as in ^1H nmr). Indeed at an early stage in the development of ^3H nmr spectroscopy a linear relationship was demonstrated between the integrated ^3H nmr signal and the specific radioactivity of a sample [4]. Subsequently it was shown that the integrated line intensities from both ^1H-coupled and ^1H-decoupled ^3H nmr spectra were the same within experimental error and that the results agreed well with those obtained from the far more time-consuming chemical and other degradation methods [20]. In fact such results (Table 9) indicated the absence of a significant differential nuclear Overhauser effect [29].

Because the tritium in tracer compounds is normally at low abundance, the ^3H nmr signals are very weak and need to be intensified by standard spectral accumulation techniques. In addition, as in ^{13}C nmr, ^1H noise decoupling is frequently applied. This concentrates into single lines the available ^3H nmr signal strength, which otherwise is spread out in multiplet signals because of coupling to the protons. However, by decoupling the protons, nuclear Overhauser effects may arise. These effects on ^3H signal intensities will be derived from the ^1H-induced

Table 9. Distribution of tritium in various organic compounds as determined by ^3H nmr spectroscopy and by degradation methods.

Compound (position tritiated)	^3H Distribution (%) [and Reference] Nmr	Degradation
$[1\beta,2\beta(n)$-$^3H]$Testosterone*	[20]	[21,22]
(1β)	40	
(2β)	39	
$(1\beta+2\beta)$		75–82
(1α)	9	
(2α)	12	
$(2\alpha+2\beta)$		50
$[1\alpha,2\alpha(n)$-$^3H]$Cholesterol†	[23]	[24]
(1α)	46	
(2α)	37	
(1β)	17	
$(1\alpha+1\beta)$		62–66
$(2\alpha+2\beta)$		40
$[G$-$^3H]$Phenylalanine*	[25]	[26]
(side chain-2)	2	4
(ring)	74	71
$[G$-$^3H]$Benzo[a]pyrene*	[27]	[28]
(6)	11	8.6
$[G$-$^3H]$Valine‡	[29]	[30]
(α)	12	10.5
(β)	74	77.5
(γ)	14	12

* In d_6-DMSO. † In $CDCl_3$. ‡ In D_2O.

relaxation of the ^3H nuclei. The irradiation of the protons increases their population of the upper spin state relative to the lower. So as to maintain an overall Boltzmann distribution between upper and lower spin states for the whole nuclear spin system, a compensating redistribution of triton spins occurs to give a greater population in their lower spin state. More radiofrequency energy can then be absorbed by the tritons, with a resultant increase in ^3H nmr signal strength. Theory [31] indicates that the maximum Overhauser effect is given by

$$\text{n.O.e.(max.)} = \gamma_X/2\gamma_Y \qquad (9)$$

where X is the irradiated nuclide and Y the observed nuclide. For ^3H signals observed with ^1H decoupling, equation (9) indicates a maximum enhancement of 47 per cent., which is not large compared with ^{13}C signal increases of up to 199 per cent. on ^1H irradiation. Indeed, the worst possible error in comparing relative intensities of ^3H signals would be ± 10 per cent. [29] but is generally much less, partly because the observed Overhauser effect for tritons is usually only 10 to 40 per cent. (Table 10) but mainly because of the lack of marked differential effects between tritons in the different positions in a labelled structure (Table 10). Consequently, the extra time (at least four times as great) involved in obtaining n.O.e.-suppressed, decoupled spectra is rarely warranted.

Table 10. ^{3}H nuclear Overhauser effects (%) with relative ^{3}H signal intensities (%) from (a) ^{1}H noise-decoupled and (b) n.O.e.-suppressed gated-decoupled spectra [29, 32]

Compound (position tritiated)	n.O.e.	(a)	(b)
[G-^{3}H]Toluene*			
(Me)	17	40.1	38.6
(*o*)	10	27.6	28.4
($m+p$)	10	32.3	33.0
[G-^{3}H]Isopropylbenzene*			
(Me)	32	26.6	25.5
(α-CH)	22	10.5	10.8
(ring)	24	62.9	63.7
[1β,2β-$^{3}H_2$]Testosterone†			
(2β)	40		
(1β)	41		
[1β-^{3}H]Testosterone†			
(1β)	24		
[1α,2α-$^{3}H_2$]Testosterone†			
(1α)	40		
(2α)	32		
[2,4,6,7(*n*)-^{3}H]Estradiol†			
(2)	19		
(4)	18		
(6α)‡	14		
(7α)‡	36		
(7α)§	37		

* Neat. † In d_6-benzene.
‡ Both 6α- and 7α-positions labelled.
§ 6α not tritiated.

7. Instrumentation

The development of ^{3}H nmr spectroscopy has been greatly assisted by improvements in instrument design and performance. In 1964 it was necessary for Tiers *et al.* [3] to use about 10 Ci of tritiated ethylbenzene in order to obtain a spectrum of satisfactory signal-to-noise ratio. During the intervening years spectrometers operating at higher magnetic fields (and so frequencies) have been brought on to the market. Table 11 contains details of the kind of instruments that are in worldwide operation by the research groups known to be using ^{3}H nmr spectroscopy. The move to higher frequencies has brought with it benefits in

Table 11. Nmr spectrometers that have been used for ^{3}H nmr analysis

Year	Details	Frequency	Reference
1959	Custom built	30	[9]
1964	Varian A-40	40	[3]
1968	Perkin Elmer R-10	64	[4]
1973	Bruker WH-90	96	[5]
1977	Varian XL-100	106.7	[32]
1978	Jeol FX-60	64	[34]
1978	Bruker WP-60	64	[35]
1980	Bruker WP-200	213.47	[36]
1980	Bruker SXP-22/100	95.9	[37]
1980	Bruker WP-100	106	[38]
1981	Bruker WM-300	320	[39]

sensitivity and increased spectral dispersion. At the University of Surrey the nmr samples usually contain between 5 and 25 mCi, a convenient range for examination at 96 MHz, but with instruments operating at the higher frequencies samples containing as little as 0.1 mCi per labelled position can be analysed. With further improvements, probably in pulse sequence design to enhance ^{3}H nmr signal strength through crosspolarization of abundant spins (e.g. ^{1}H) in the sample [33], as well as in instrument performance, future trends may not be too far away when samples at truly tracer level, i.e. a few microcuries, can be analysed by ^{3}H nmr spectroscopy.

8. Sensitivity and limits of detection in ^{3}H nmr spectroscopy

Some points have of necessity already been made, so the following is a summary. It has already been mentioned (see Table 1 and page 2) that as a consequence of their high nuclear magnetic moment, tritons have an improved receptivity or sensitivity to nmr detection, by the factor 1.21, $(\gamma_T/\gamma_H)^3$, as compared with protons in equal numbers at the same magnetic field strength. In the routine analysis of tritiated compounds, prepared on a commercial scale, there is normally an adequate amount of material available (tens of millicuries) so that acquisition times are reasonable (see Figure 5). However, this is not necessarily so in many *applications* of tritiated compounds. Here it is often necessary to analyse tritiated compounds not only at low isotopic abundance but also at low concentration. The superior receptivity of the triton clearly helps but it is often necessary to use ^{1}H noise decoupling (page 13) and to acquire the ^{3}H spectrum over a prolonged period. For this to succeed there must be sufficient intrinsic sensitivity and long term stability in the nmr spectrometer equipment. The proper matching of the level of labelling to the capability of the instrumentation is essential (see Figure 5) [40]. For very weakly radioactive samples (< 1 mCi) a higher field nmr spectrometer would be required to take advantage of the resultant increase in signal-to-noise ratio which improves as the 3/2 power

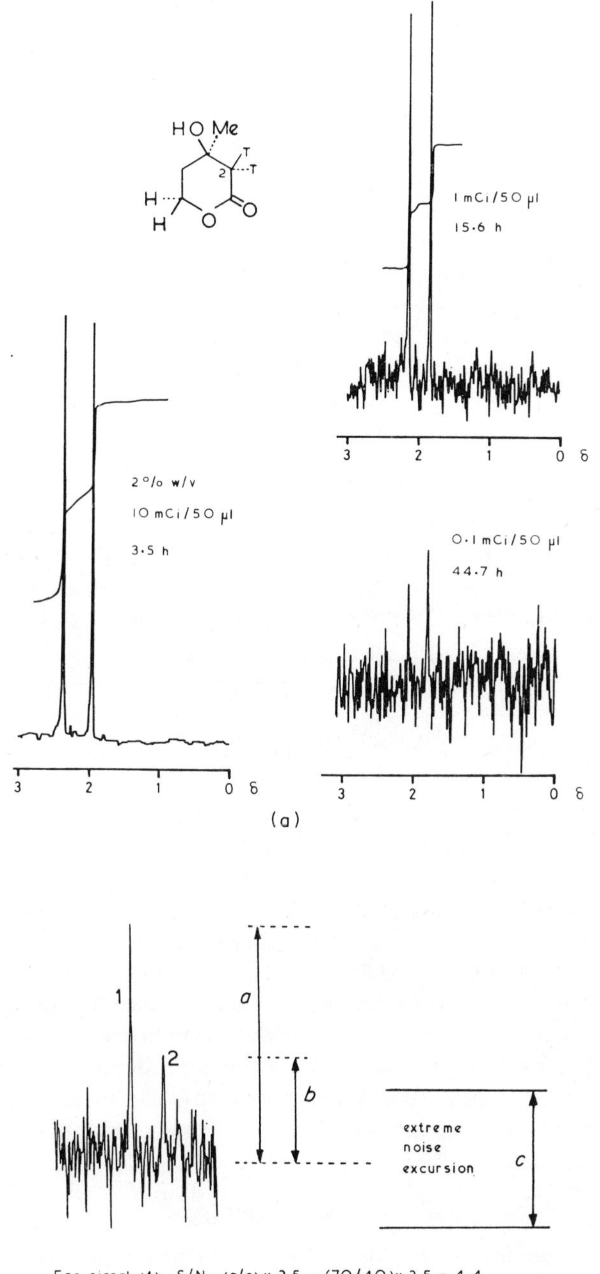

Figure 5. (a) ^{3}H nmr spectra (^{1}H decoupled) of [2-^{3}H]mevalonolactone in d_6-benzene at successive dilutions, acquired with quadrature detection at 96 MHz; (b) illustration of measurement of signal-to-noise ratio

of the spectrometer operating frequency [41]. Thus if a tritiated compound bearing 10 mCi (0.37 GBq; 0.035 per cent. isotopic abundance) per site gives an adequate ^{3}H nmr spectrum with an FT spectrometer operating at 64 MHz (1.4 T), then at 96 MHz (2.11 T), 5 mCi would suffice, whilst with a 9.38 T supercon FT spectrometer (426.66 MHz), only $10 \times (64/426.66)^{3/2} = 580\,\mu$Ci would be needed, if the acquisition times were kept about the same. Because the signal-to-noise ratio (S/N) improves as $t^{1/2}$ where t is the spectral acquisition time, increasing the acquisition time four times with the most powerful of these spectrometers would allow the ^{3}H nmr examination of as little as 200–300 μCi per site, which is 0.0007 to 0.001 per cent. tritium per site (cf. page xi and [39]), the maximum specific radioactivity (given by 100 per cent. tritium labelling) at one site being 28.76 Ci mmol^{-1} (Table 2). The S/N ratio, defined as S/N = (signal height/extreme noise excursion) × 2.5, is measured as indicated in Figure 5b.

9. Setting up the spectrometer

Partially monotritiated dimethyl sulphoxide (DMSO) is an ideal compound to use as a test sample for setting up a Fourier transform nmr spectrometer for ^{3}H resonance studies [42]. About 20 to 30 mCi will provide sharp resonance lines after a few transients. The [^{3}H]DMSO is readily prepared by exchange [18], is then diluted with redistilled perdeuterated dimethyl sulphoxide and sealed as described in Section 10 (below). Being self-scavenging for radicals, the sample is radiolytically stable over long periods (at least 2 years) and no detectable further exchange occurs. The field homogeneity and lock are adjusted from the ^{2}H nmr signal as usual. With the pulse point set close to the ^{3}H nmr signal and with adequate ^{1}H noise decoupling, the acquisition parameters can readily be adjusted to give a free induction decay signal which appears on the monitor as a well shaped, low frequency sine wave. A few transients will then transform to a properly shaped singlet line. Optimization of the ^{1}H decoupling power is also facilitated because the test sample will give a proton-coupled ^{3}H nmr signal which is a relatively wide triplet ($^2J_{TH}$ = 13.2 Hz). If a spectrometer is used which has a fluorine-19 channel for field-frequency locking, the test sample would need to be diluted with an equal volume of a miscible fluorocompound. Direct comparison of ^{3}H with ^{1}H nmr spectra can be facilitated by using identical p.p.m. scales, viz. 21.33 Hz cm^{-1} for ^{3}H and 20.00 Hz cm^{-1} for ^{1}H (ratio, ν_T/ν_H) or appropriate equal fractions or multiples of these.

10. Sample preparation

Spherical microbulbs of the type described by Flath, Henderson, Lundin and Teranishi [4, 43] were used in the early ^{3}H nmr studies (Figure 6a). These had the advantages that (a) the sample volume is small ($\sim 30\,\mu$l) so that the total radioactivity is kept as low as possible (compatible with the sensitivity of the instrument); (b) the microbulbs are extremely strong and resist, for example, being dropped on a linoleum floor; (c) the bulbs withstand surprisingly high

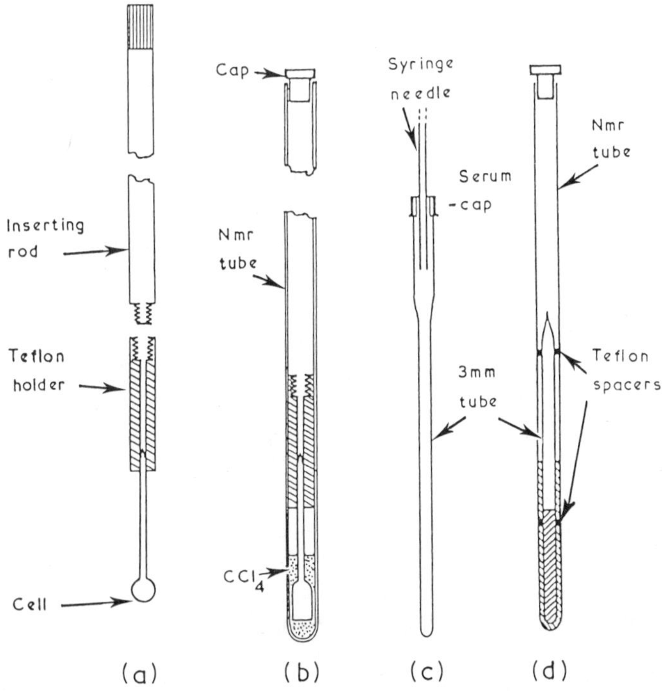

Figure 6. Details concerning microcells for ^3H nmr spectroscopy: (a) spherical cell, chuck and insertion tool; (b) nmr tube containing cylindrical microcell; (c) preferred 3 mm tube, as for filling; (d) 3 mm sample tube mounted in 5 mm nmr tube.

internal pressure, which might develop as a result of self-radiolysis—samples having a radioactivity of 10 to 50 Ci ml^{-1}, 0.37 to 1.85 TBq ml^{-1}, have been kept for up to 6 months in such bulbs without breakage, but no more than 10 to 50 mCi ml^{-1} needs to be handled in general, and at this level self-radiolysis is not perceptible (by nmr) during months; and (d) with the microbulb mounted inside the nmr tube the sample is doubly protected and therefore the assembly is very safe in use. Subsequently we used 100 μl cylindrical cells, with a useful gain in sensitivity. These microbulbs are best loaded by means of a 'Hamilton' 100 μl syringe. Volatile compounds are loaded, using a refrigerated syringe and bulb, in a dry-box. The microbulb is carefully sealed so that it fits the Teflon holder (Figure 6a, b) and will slide properly into the nmr tube without strain, carbon tetrachloride being placed in the tube so as to surround the bulb to a depth of ~ 3 cm: this minimizes wobble when the complete assembly is spun in the spectrometer probe. The microbulb is slid into the nmr tube to a predetermined position, for optimum signal strength, the threaded insertion tool is then withdrawn and a cap is placed on the tube (as in Figure 6b).

With experience it became clear that spherical microbulbs had two disadvantages. Firstly, the narrow capillary necks made introduction of the sample difficult. Secondly, air bubbles would sometimes be trapped in the microbulb and these would adversely affect the instrument resolution. A tubular microcell assembly (Figure 6c) with a 5 mm filling extension was then developed so as to overcome these difficulties [44]. The extension is fitted with a standard nmr serum cap, enabling samples to be handled in the absence of air and sealed under inert gas or in a partial vacuum. Most frequently, the serum-capped tube assemblies are evacuated via a syringe needle; they maintain the partial pressure for days. Filling by syringe through the cap is then considerably facilitated, as also is the flame-sealing of the neck, below the 5 mm section, to leave a precision 3 mm tube containing 50 μl of sample solution about 1.5 cm in depth. The tubular cell or ampoule is then mounted in slotted Teflon rings (precisely machined) and slid into the 5 mm nmr tube to a predetermined position, and the tube is capped (Figure 6d). Usually the sample has been dissolved in a deuterated solvent to provide the deuterium nmr signal for the time-shared, field-frequency lock, and the appropriate internal ^{1}H reference (e.g. TMS or DSS) has been added. When a deuterated solvent is not employed for the sample, then the annular space between the 3 mm ampoule and the 5 mm nmr tube is filled with a deuterated liquid to provide an external lock, whilst the ^{1}H reference will have been added as usual to the sample solution itself.

For samples of limited solubility or low specific activity it may be necessary to increase the sample size by use of 300 or 500 μl cylindrical bulbs in 10 mm nmr tubes (cf. Figure 6b). The increase in size may bring problems of poor spectral resolution, dependent upon magnet homogeneity. Moreover, the larger bulbs are not as rugged as the smaller type so that on the whole they are infrequently used.

Calvert, Martin and Odell [34] have described work done with a 450 μl bulb system on a Jeol FX-60 instrument and in 1978 Sawan and James [45] described the use of sealed plastic capsules which, they claimed, would remove hazards from breakage. However, their effective sealing demands skill, and in view of the well known incompatability of some solvents and organic compounds with polyethylene (and other plastics), there may be some restriction in their use.

Tritiated compounds derived from tracer investigations are usually in very dilute solution (micrograms per millilitre) and normally require concentration by lyophilization (freeze-drying) before dissolution in a small volume of nmr solvent containing internal ^{1}H reference. Occasionally, lyophilization can result in decomposition of sensitive compounds. Heat, light, traces of metal ions, traces of enzymes from the skin and incorrect pH can all seriously affect the stability of some classes of compound especially tritiated nucleoside triphosphates, for example, and extreme care is needed in the preparation of such samples. This is apart from the necessary precautions for handling radioactive materials.

11. Safety

The general health physics aspects of working with radiotracers are adequately described by Muramatsu *et al.* [46] and for tritium labelled compounds in particular by Evans [1]. Tritium is one of the least toxic radionuclides but even so care is still required in the handling of tritium labelled compounds for ^{3}H nmr spectroscopy, especially to avoid contamination of the laboratory. Although the levels of radioactivity involved in ^{3}H nmr spectroscopy are higher (an order of magnitude or so) than those customarily associated with work in a low-level tracer radiochemistry laboratory no new facilities are needed and the general safety precautions necessary are very much the same, as indicated in the summary in Table 12. When the samples are sent for ^{3}H nmr analysis it is important that neither the instrument operator is placed at risk nor the laboratory contaminated. Thus the capped nmr tubes should be carried in a metal can containing Vermiculite. The can should bear proper identification and hazard labels, and the lid must be taped on. Transport of samples to the nmr instrument laboratory is then completely without hazard. At the spectrometer, the double glass containment of the samples makes their handling very safe. In fact no breakage of tritium samples or contamination has occurred in the nmr laboratory of the University of Surrey during the handling of some thousands of samples over the past 15 years.

Table 12. Safety precautions in the handling of tritiated compounds

1. Confine all radioactive work to the radiochemical laboratory, segregating clean from contaminated apparatus.
2. Always use the minimum quantity of radioactivity compatible with the objectives of the experiment.
3. Always work over a spill tray and in a ventilated enclosure if the general ventilation of the laboratory is not appropriate.
4. Whilst in the radiochemical laboratory always wear clean protective clothing, safety glasses and rubber gloves; these should be removed before leaving and the hands then washed.
5. Respect and rigidly conform to the Laboratory Rules for segregation, storage, use etc., of the safety clothing and protective equipment.
6. Always work carefully and tidily taking care to label containers carrying tritiated compounds, giving details of the compounds, amount, specific activity and date prepared. Never pipette radioactive materials by mouth.
7. Do not eat, drink, apply cosmetics, or smoke in the radiochemical laboratory. Use paper tissues rather than ordinary handkerchiefs, and dispose of them as if they were active low-level waste.
8. Never work with cuts or breaks in the skin unprotected.
9. In the event of a spillage it is essential to minimize the spread of contamination. Identify the contaminated area and starting from the outer edge, decontaminate the area in convenient sectors by wiping and scrubbing, followed by monitoring for surface contamination.
10. Arrange for all radioactive waste to be placed in clearly labelled containers and dispose of it according to statutory requirements.
11. Arrange for the regular monitoring of personnel by the collection and radioactive counting of urine samples.

Monitoring of bench tops, instruments and other laboratory equipment should be carried out frequently, by swabbing surfaces with moist cotton wool or paper, which is then counted in a liquid scintillation counter. When performed on a regular basis, especially after each usage of millicurie amounts of tracer, this method provides a good and relatively inexpensive way of checking for possible surface contamination. A simple and inexpensive method for checking for airborne tritium contamination is to draw air through a trap containing liquid scintillator solution, which in turn is monitored for radioactivity in a liquid scintillation counter.

Personnel should be monitored regularly on a 2 week schedule, and additionally during and after the handling of particularly large amounts of radioactivity (tens of millicuries) or of volatile tritiated compounds. This not only provides assurance to the operator but also acts as a check on the facilities and operator care. Although measurement of tritium in breath samples is possible [1, 47], it is rather a time-consuming procedure and urine sample analysis is usually preferred. The biological half-life of tritium (as tritiated water—the most likely ingested form) in adult humans is about 12 days, and urine samples voided at a frequency appropriate to the degree of radioactivity handled provides a convenient method for the analysis of tritium uptake. To avoid problems, 'good housekeeping' must be the rule of the day.

CHAPTER 2

Labelling patterns and methods for tritium labelling

A detailed review of the methods which are available for the preparation of tritium labelled compounds was published by Evans in 1974 [1]. This chapter therefore includes an updated review of the methods currently employed; they fall into three main categories:

Chemical syntheses
Hydrogen–tritium exchange reactions
Biochemical methods

In all these cases, tritium nuclear magnetic resonance spectroscopy serves to reveal the labelling pattern conveniently and rapidly. Within the above three categories, there are those methods which are expected to lead to specific labelling in molecules and those methods which tend to lead to more general, unpredictable, labelling patterns. In practice, even those methods expected to give specific labelling may lead to non-specific labelling patterns: this is because metal/hydrogen (tritium) transfer catalysts are frequently involved.

1. Chemical syntheses

(a) Specific labelling methods

There are four principal methods used for specific labelling of organic compounds with tritium. These are:

(i) Reduction of unsaturated compounds with tritium gas using platinum or palladium metal catalysts. This category includes the reduction of alkynes, alkenes, aldimines, etc. An example is the specific catalytic reduction of methyl stearolate to methyl [9,10-^{3}H]oleate:

$$CH_3(CH_2)_7C{\equiv}C(CH_2)_7CO_2CH_3 \xrightarrow[Pd]{T_2} CH_3(CH_2)_7CT{=}CT(CH_2)_7CO_2CH_3$$

(ii) Catalytic halogen–tritium replacement reactions, e.g. the replacement of chlorine, bromine or iodine in substituted aromatic compounds.

$$p\text{-}BrC_6H_4CH_2CH(NH_2)COOH \xrightarrow[Pd]{T_2} p\text{-}TC_6H_4CH_2CH(NH_2)COOH$$

Thus tritium in the presence of palladium catalysts (supported and unsupported) will hydrogenolyse the C–Br bond in *p*-bromophenylalanine to give [4-^{3}H]phenylalanine. The method is less satisfactory for labelling aliphatic and alicyclic compounds.

(iii) Reduction of functional groups such as –CHO, >C=O, $-CO_2R$, –CH=NH with tritiated metal hydrides. Lithium, sodium or potassium borotritides, sodium cyanoborotritide and lithium aluminium tritide are all employed, in particular circumstances dictated by solubility, stability of substrate, solvent and specificity of labelling considerations. For example, aldehydes yield [1-^{3}H]alcohols after reduction by sodium borotritide and work-up with water or hydroxylic solvents:

$$\text{R–CHO} \xrightarrow{NaBH_3T} \text{R–CHTOH}$$

(iv) Reactions involving tritiated methyl iodide. Tritiated methyl alcohol can be made at 100 per cent. isotopic abundance, by catalytic reduction of carbon monoxide (or dioxide) with tritium in a Fischer Tropsch type synthesis [48]:

$$CO_2 + 3T_2 \xrightleftharpoons[\substack{200\text{–}250\,^\circ C \\ 150\text{–}600\text{ p.s.i.}}]{\substack{Cu/Zn/Cr \\ \text{mixed catalyst}}} T_2O + CT_3OT \qquad (10)$$

Subsequent conversion into [^{3}H]methyl iodide with hydriodic acid provides an excellent intermediate for numerous compounds labelled in the methyl group. Such syntheses can involve simple *N*-methylation, as, for example, in the synthesis of [*N-methyl*-^{3}H]diazepam [19, 49]:

$$\text{(N-H benzodiazepinone, Cl, Ph)} \xrightarrow{CT_3I} \text{(N-}CT_3\text{ benzodiazepinone, Cl, Ph)}$$

Alternatively, the [$^3H_3$]methyl iodide can be used in more complex reactions as in the preparation of 25-hydroxy[26,27-*methyl*-^{3}H]cholecalciferol [(3*β*, 5*Z*, 7*E*)-9,10-secocholesta-5,7,10(19)-trien-3-ol] from [$^3H_3$]methylmagnesium iodide and 25-oxo-26-norcholecalciferyl 3-acetate [50]:

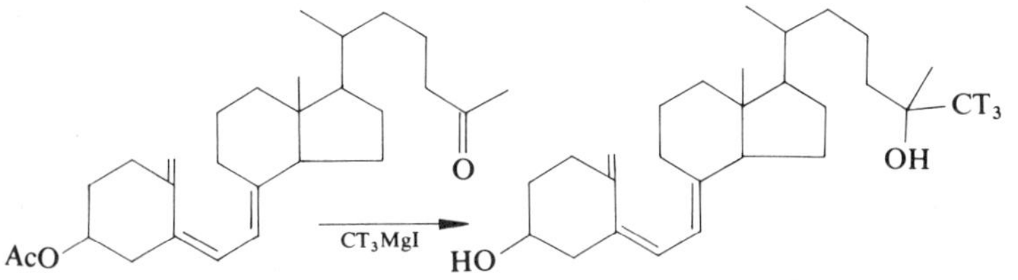

AcO CT$_3$MgI HO OH CT$_3$ O

Many of these general synthetic methods for tritium labelling do in fact lead to the high degree of specific labelling expected. In any case it is possible to select reaction conditions so that the degree of non-specific labelling is kept to a minimum (say less than 10 per cent). However, in the catalytic methods, especially when using the platinum group metals, tritium can be incorporated at random. This often occurs in the reduction of ethylenic double bonds over platinum or palladium catalysts, where double bond migration followed by saturation with tritium can occur. The classical example of non-specificity in catalytic addition of tritium to a carbon–carbon double bond is the preparation of [^{3}H]stearic acid by the platinum catalysed hydrogenation of elaidic acid (*trans*-9,10-octadecenoic acid) with tritium gas. Subsequent analysis by chemical degradation showed that only 15 per cent. of the incorporated tritium was actually attached to the 9 and 10 carbon atoms. The other 85 per cent. was distributed along the carbon chain, so that 37.3 per cent. of the total tritium was attached to carbon atoms 1 to 9 with the remaining 62.7 per cent. attached to carbon atoms 10 to 18 [51]. Fortunately this is rather an extreme example of non-specific labelling with tritium by chemical synthesis.

Another example illustrates and draws attention to the dangers of relying upon chemical degradation methods for establishing the patterns of tritium labelling, reinforcing the benefits of the direct and non-destructive ^{3}H nmr spectroscopic technique. In this example, folic acid (**III**, with R=OH, R′=H, X=H) and methotrexate (**III**, with R=NH_2, R′=Me, X=H) were labelled, using tritium gas, by palladium catalysed halogen–tritium replacement, starting respectively with 3′, 5′-dibromofolic acid (**III**, with R=OH; R′=H, X=Br) and 3′,5′-dichloromethotrexate (**III**, with R=NH_2, R′=Me, X=Cl). The positions of the label in the [^{3}H]folic acid were investigated by oxidative degradation with alkaline permanganate. This cleaves the C^9–N^{10} bond to give the pteridine-6-carboxylic acid (**IV**, with R=OH or NH_2, R″=COOH) and 4-aminobenzoylglutamic acid (**V**), the tritium originally in the 9-methylene group appearing as tritiated water under the oxidative conditions.

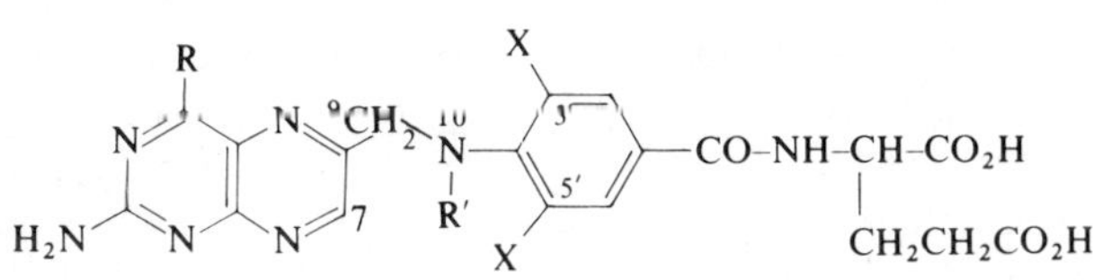

R X N $^9CH_2$ 10 N 3′ CO–NH–CH–CO_2H N 5′ H_2N N N 7 R′ X $CH_2CH_2CO_2H$

(III)

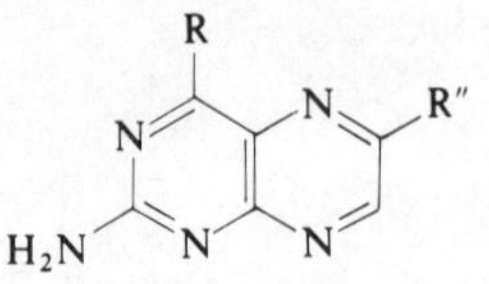

(IV)

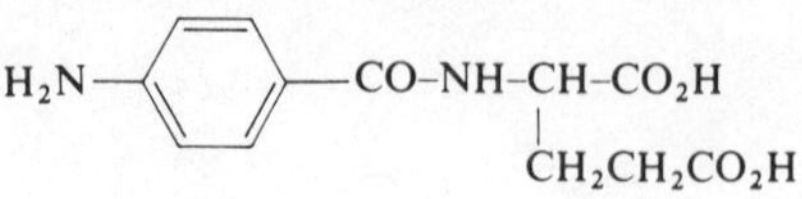

(V)

From initial degradative experiments [52] both organic fragments were radioactive. Evidently, in addition to the 3′,5′-positions of folic acid bearing tritium, tritium was present also in the 7-position of the pteridine ring. From the radioactivity of the water, it appeared that 40 per cent. of the total tritium was at the 9-methylene position. Studies by ^{3}H nmr spectroscopy [53] showed a considerably lower percentage of tritium in the 9-methylene group of folic acid (and none in the 9-methylene of methotrexate) than found by measurement of tritiated water from the permanganate oxidation. A summary of these results is shown in Table 13.

Table 13. Labile tritium by oxidation of tritiated folic acid and methotrexate. (Reproduced by permission of John Wiley & Sons Ltd.)

Compound	Distribution of label by ^{3}H nmr spectroscopy (%)			Labile tritium after oxidation, assumed from the 9-position (%)
	3′,5′	7	9	
[7,9-^{3}H]Folic acid	0	59	41	42.4
[3′,5′,7,9-^{3}H]Folic acid	42.5	25.5	32	41.8
[7-^{3}H]Methotrexate	0	100	0	8
[3′,5′,7-^{3}H]Methotrexate	43	57	0	21.2
[7-^{3}H]Pteridine-6-carboxylic acid	0	100	0	8
4-Amino-[3,5-^{3}H]benzoic acid	100	0	0	100

The discrepancy between the results was due to tritiated water arising during the permanganate oxidation, from the 3′,5′-positions as well as from the 9-position. The real distribution of tritium in labelled folic acid and in methotrexate is clearly indicated by the ^{3}H nmr spectra shown in Figures 7 and 8, respectively.

Tritiated folic acid and its 4-amino-N^{10}-methyl analogue, methotrexate, may be used in biomedical experiments where the possibility of metabolic alteration of the compound exists. It is then very important that the correct original distribution of the label should be known. The method of synthesis suggests that the tritium would be in the 3′,5′-positions only. However, the tritium nmr spectrum of the [^{3}H]folic acid (Figure 7) at once shows that only 42.5 per cent. of the total tritium is in the 3′,5′-positions with 25.5 per cent. located at the

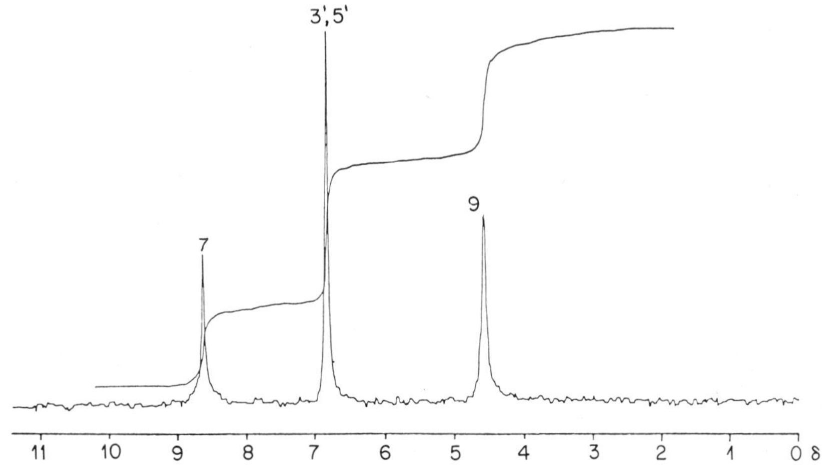

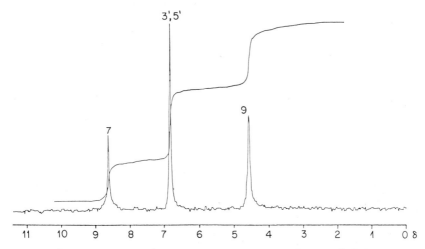

Figure 7. ^{3}H Nmr spectrum (^{1}H-decoupled) of potassium [3′,5′,7,9-^{3}H]folate in D_2O. (Reproduced by permission of John Wiley & Sons Ltd.)

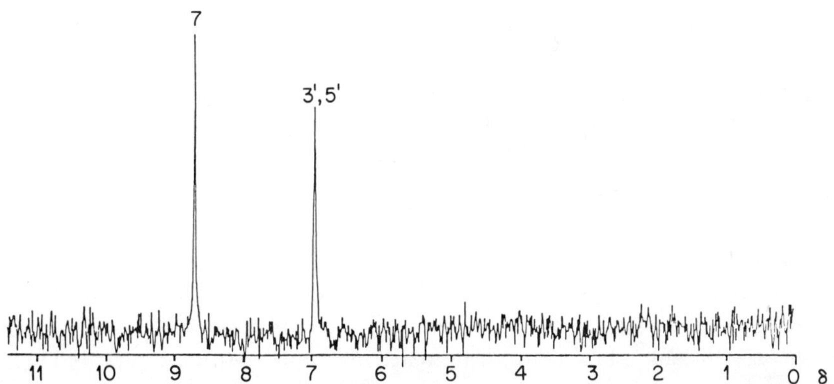

Figure 8. ^{3}H Nmr spectrum (^{1}H-decoupled) of sodium [3′,5′,7-^{3}H]methotrexate in D_2O. (Reproduced by permission of John Wiley & Sons Ltd.)

7-position and 32 per cent. in the 9-methylene group. For methotrexate labelled by the same procedure the ^{3}H nmr spectrum (Figure 8) shows that the 3′,5′- and 7-positions are labelled, but that the 9-methylene position is unlabelled, probably because of steric hindrance by the N^{10}-methyl group. As one might expect from an exchange reaction, the degree of labelling in each position will vary from batch to batch (see later, Table 30).

When the aim is to produce specifically tritiated compounds, it is obviously most important to use starting materials of the highest purity. This does not mean just compositional purity but structural homogeneity. When samples of 6-bromobenzo[a]pyrene were subject to tritio-debromination [27], the products from different batches showed differing ^{3}H nmr spectra (see Figure 9) indicative

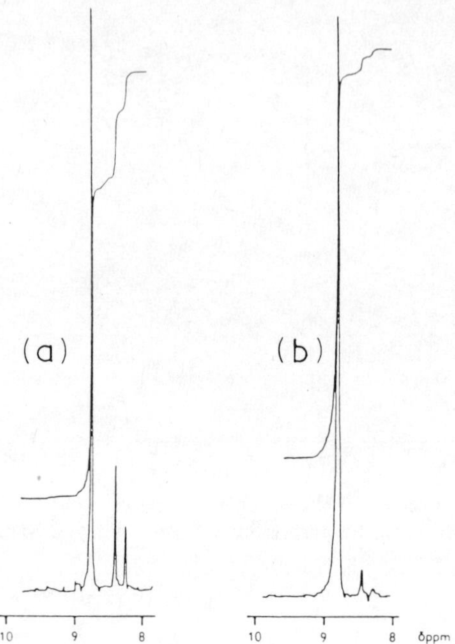

Figure 9. ^{1}H-Decoupled ^{3}H nmr spectra of [6(n)-^{3}H]benzo[a]pyrene (in d_6-DMSO) prepared from two different samples of 6-bromobenzo[a]pyrene. (Reproduced by permission of John Wiley & Sons Ltd.)

of significant, varying amounts of tritium at the 1- and 3-positions in addition to that required in the 6-position. For sample (a), the ^{3}H spectrum showed 15 per cent. of tritium at the 1-position and 4 per cent. at the 3-position, whilst for sample (b) there was 5 per cent. at the 1-position and 3 per cent. at the 3-position. These results were found to be due to the presence of 1- and 3-bromo compound in the 6-bromobenz[a]pyrene and not to tritium migration or adventitious catalysed hydrogen–tritium replacement.

Catalytic tritiodebromination of 7-bromopregnenolone also gave other than the expected tritiated product but not for the same reason. The ^{3}H nmr spectrum of the product taken with ^{1}H decoupling (Figure 10) exhibited three singlets. Those at $\delta = 1.49$ and 1.90 with relative intensities of 52 and 28 per cent. arose from tritium at the 7α- and 7β-positions, as expected. The third signal at $\delta = 2.13$, with intensity 20 per cent., could be assigned to the 4-position. The hydrogen at this position is allylic and so is very likely to have undergone exchange under the reaction conditions: there was no evidence that this position had been brominated. In the undecoupled ^{3}H spectrum, the signal at $\delta = 2.13$ showed a major doublet splitting of approximately 15 Hz, consistent with geminal (T, H) coupling. There was no additional large axial–axial coupling (as there would be between 4β-H and 3α-H) and so the 4α-assignment for the extra signal was confirmed [20].

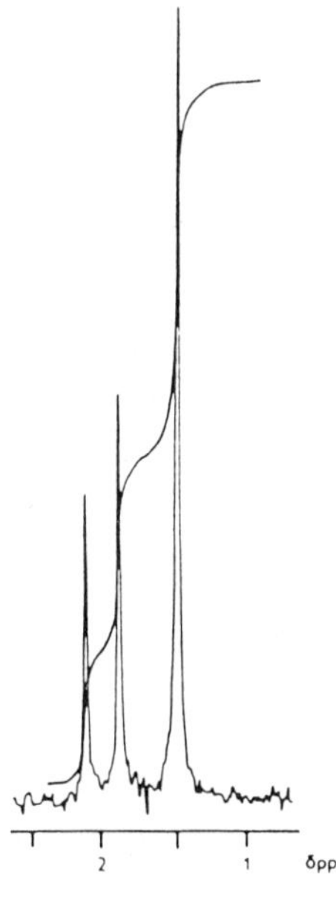

Figure 10. ^{3}H Nmr spectrum of [7(*n*)-^{3}H]pregnenolone with ^{1}H-decoupling) in d_6-DMSO. (Reproduced from *Steroids*, **28**, p. 371 (1976) by permission of Holden-Day Inc.)

The reduction of functional groups with tritiated metal hydrides is normally highly specific and results in an insignificant amount of non-specific labelling. This is discussed in the section on labelled carbohydrates (see page 43).

Similarly, the tritium in the methyl group of [^{3}H]methyl iodide remains firmly bound during further synthesis, and compounds prepared from this intermediate are specifically labelled. Thus L-[*methyl*-^{3}H]methionine, prepared from [^{3}H]methyl iodide and homocysteine (from *S*-benzylhomocysteine), has a specific activity equal to that of the starting methyl iodide, with all the tritium located in the methyl group [19, 54].

A further example of non-specific labelling in chemical synthesis serves also to draw attention to the use of ^{3}H nmr spectroscopy in eliminating uncertainties concerning the metabolic stability of the label. At present the use of tritiated compounds for pharmacokinetics and metabolism studies is of less importance than the use of [^{14}C]compounds for such studies. Nevertheless, in cases where the carbon-14 labelling is impracticable or technically very costly, tritium labelling may well be considered, provided that the tritium atoms can be introduced into positions which are metabolically stable. The possible biological stability of the label in a compound can often be judged from known general biodegradation pathways and from the metabolic fate of structurally related compounds. An appropriate intermediate for tritiation can then be selected, and by using ^{3}H nmr spectroscopy the distribution of tritium in the labelled product can readily be checked before use or before the tritiated compound's biological stability is tested by direct experiment. CU32-085 (**VI**) is a highly potent inhibitor of prolactin secretion which could conceivably be tritiated via reduction of either of the two isomeric $\Delta^{7,8}$- and $\Delta^{9,10}$-lysergic acid methyl esters (**VII** and **VIII**,

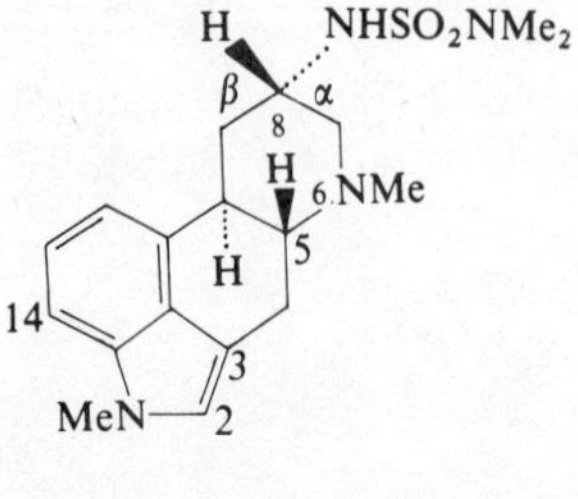

(VI) 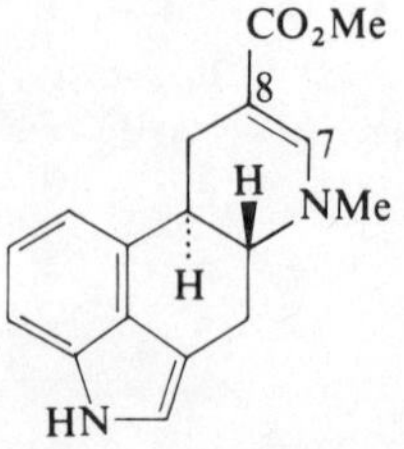 (VII) 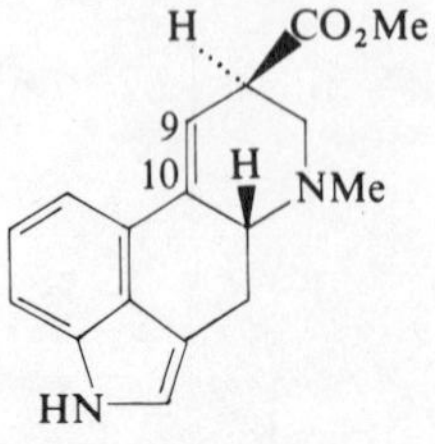 (VIII)

respectively). Reduction of the 7,8-position double bond with tritium unexpectedly gave a product of low specific activity, but a more serious setback was the finding by ^{3}H nmr that 42 per cent. of the tritium was located at the 2-position and 25 per cent. at the 14-position, both these positions being susceptible to biotransformation processes. In contrast, when the 9,10-position double bond ester (**VIII**) was reduced with tritium, 97 per cent. incorporation was achieved in the expected positions. Although the subsequent steps to form tritiated **VI** were more demanding than if the route from **VII** had been used, the tritium hydrogenation results left no alternative [55].

2. Hydrogen–tritium exchange reactions

Just as methods for specific labelling with tritium can give rise to a degree of non-specific labelling, so general labelling methods can sometimes do the opposite and give rise to specific labelling. However, neither of the positions labelled nor the quantitative distribution of tritium in a molecule can be predicted with certainty: this can only be determined experimentally, and here ^{3}H nmr spectroscopy is ideal. Methods for general labelling with tritium have been well described and their advantages and disadvantages discussed in detail [1]. General labelling is achieved by exchange reactions either by use of tritium gas or a tritiated solvent, often with catalysts.

(a) Use of tritium gas

Tritium–hydrogen isotope exchange in organic compounds induced by exposure to tritium gas was first recognized by Wilzbach [56] in 1956. Generally, there is much radiation damage to the starting material, with the production of a range of different, highly tritiated products. Although there have been numerous modifications to this radiation-catalysed exchange procedure, the continuing problems of purification of the required [G-^{3}H]product, and the relatively low molar specific activities normally achieved, have made the method of limited use. Examples for which successful Wilzbach labelling has been achieved include

L-[^{3}H]proline (page 40) and [^{3}H]atropine (page 148) where 60 to 75 per cent of the tritium introduced reproducibly goes into specific positions.

A more widely used method involving tritium gas is that developed by Evans *et al.* [57], based on the ability of hydrogen atoms in certain positions in molecules in solution to exchange with tritium gas in the presence of a metal catalyst (usually palladium). Carbohydrates, amino acids, purines, nucleosides, nucleotides and steroids have all been labelled by this method [58], which yields tritiated products of high specific activity and purity, and often with specific labelling (see the tables in Section 4). An extension to the tritiation of physiologically active polypeptides has recently been reported [59].

Similarly, dibenzyl has been labelled by exchange with tritium gas in dioxan using a reduced palladium oxide catalyst [60]. The nmr spectrum of the corresponding deuterated compound, prepared under the same conditions, showed specific labelling in the methylene groups, in agreement with the work of Evans *et al.* [57].

(b) Use of tritiated solvents

The most widely used of the methods for obtaining generally labelled compounds involves heating the compound (usually in a sealed tube and to 120 to 180 °C) for a few hours (5 to 20 h) in tritiated water, acetic acid, dimethylformamide or other solvent of total activity 5 to 500 Ci ml^{-1} in the presence of a metal catalyst prepared by reduction in hydrogen. A wide range of heterogeneous catalysts (Pt, Pd, Ni, Co, Fe, Rh, etc.) in supported or unsupported forms have been employed, platinum being the most active. Compounds labelled by this procedure include, for example, amino acids, polycyclic aromatic hydrocarbons, purines, pyrimidines and steroids [1].

Many organic compounds can also be labelled specifically under homogeneous conditions by taking advantage of the weakly acidic character exhibited by some carbon–hydrogen bonds [61]. Thus, in the presence of a suitable base, specific ionization can occur allowing hydrogen isotope exchange with the appropriately tritiated solvent. Some compounds such as the β-diketones are sufficiently acidic to permit labelling to take place even in neutral solution at ambient temperature. For weaker acids such as ketones, strong bases such as the hydroxide ion are required to produce the intermediate carbanion. For compounds that are even less acidic, such as triphenylmethane or toluene, highly basic media (e.g. alkoxide in dimethyl sulphoxide) are necessary for exchange to take place.

Treatment of many organic compounds with strong acid results in protonation. Subsequent proton loss, if it involves other hydrogen atoms in the compound, can provide a means of exchange. All the hydrogen atoms in benzenoid and polycyclic aromatic hydrocarbons may be exchanged in this way. For highly complex molecules, some specificity may result. Thus treatment of the unstable, light sensitive, antitumour drug vincristine (**IX**) with [*carboxyl*-^{3}H]-trifluoroacetic acid at ambient temperature, followed by back exchange of labile tritium with methanol, provided, as with vinblastine [62], a tritiated drug of high specific activity suitable for development of a radioimmunoassay. In this case the ^{3}H nmr spectrum showed that the label was confined to the stable aromatic positions in the indole ring (Figure 11), in contrast to vinblastine where the 17 position was also labelled.

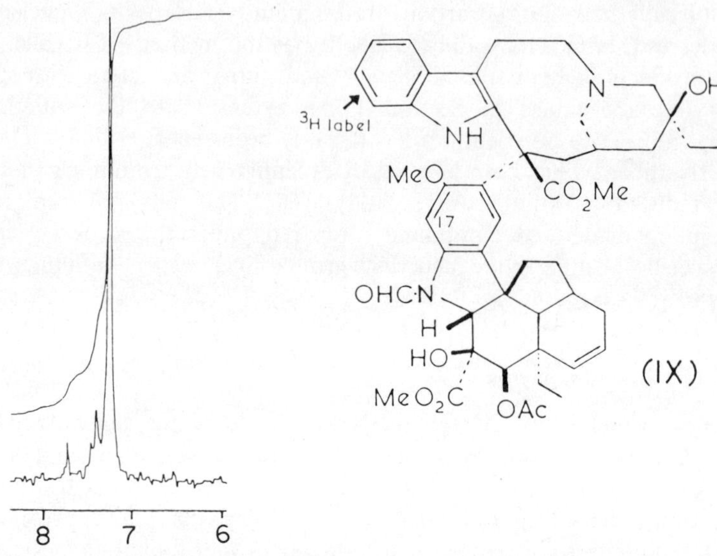

Figure 11. ^{1}H-Decoupled ^{3}H nmr spectrum of [^{3}H]vincristine sulphate in D_2O. (Reproduced by permission of D. Reidel Publishing Company.)

Numerous such studies of acid-catalysed tritiations have been reported [1, 63 to 65]. In extending the method to deuteration of aromatics the use of high temperature and low acid concentrations have been favoured for some compounds [66]. A number of Lewis acids have also been used to induce hydrogen isotope exchange, and of these ethylaluminium dichloride [67, 68] is amongst the most reactive. The process involves formation of a substrate–Lewis acid complex and then decomposition of this with tritiated water. In all exchange reactions prerequisites are that the substrate (a) is of extremely high purity (impurities may become labelled in addition to the substrate) and (b) is reasonably stable under the experimental conditions. The patterns of labelling in tritiated compounds prepared by catalysed hydrogen–tritium exchange are further discussed in Chapter 3 under studies of catalysis.

3. Biochemical methods

Occasionally, biosynthetic methods are more attractive for the preparation of particular compounds. When such compounds are required in a labelled form, then a tritiated precursor is employed in the biosynthesis [1], prepared either by a specific or a general tritiation procedure, as appropriate. These biochemical methods usually involve the use of purified or partially purified enzymes and unless the enzymes act at or involve the actual positions labelled, then the original pattern of tritium labelling is preserved in the derived products. Thus the enzymatic conversion of [*methyl*-^{3}H]thymine to [*methyl*-^{3}H]thymidine, and of L-[*methyl*-^{3}H]methionine to *S*-adenosyl-L-[*methyl*-^{3}H]methionine, for example, occur without any detectable change in the labelling patterns. On the other hand, the loss of tritium from the 2-position of DL-[G-^{3}H]amino acids on treatment with D- or L-amino acid oxidases, due to the presence of contaminating transaminases, was amongst one of the earliest examples of bioinstability of a tritium label [1, 69]. In such cases the conversion of the tritiated substrate results in a different distribution pattern in the product. Another example is in the resolution of DL-[2,3-^{3}H]aspartic acid, using hog kidney acylase-III on the *N*-acetyl derivative. Tritium nmr spectroscopy showed that whilst there was effectively no change in the distribution pattern in converting DL-[2,3-^{3}H]aspartic acid into its *N*-acetyl derivative with acetic anhydride, acylase treatment gave L-[2,3-^{3}H]aspartic acid with a much diminished amount of tritium in the 2-position. The results [23] are summarized in Table 14.

Table 14. Tritium distribution in [^{3}H]aspartic acid and its *N*-acetyl derivative, and after treatment of the latter with hog kidney acylase-III to give L-[^{3}H]aspartic acid.

Compound	Position	Chemical shift δ	Tritium (%)
DL-[2,3-^{3}H]Aspartic acid	2	3.84	33
	3	2.76	31
	3	2.63	36
N-Acetyl-DL-[2,3-^{3}H]aspartic acid	2	4.35	33
	3	2.65	35
	3	2.46	32
L-[2,3-^{3}H]Aspartic acid	2	3.86	12
	3	2.75	35
	3	2.69	53

Such changes in the tritium distribution patterns in tritium labelled compounds are very easily observed by ^{3}H nmr spectroscopy, making this is an excellent technique for biochemical reactions, as will be seen from examples in Chapter 3.

4. Labelling patterns

Since 1968 more than 3,000 samples of tritium labelled compounds have been examined by ^{3}H nmr spectroscopy at the University of Surrey in order to determine the distribution of tritium. The results for a number of classes of compounds are now discussed, covering especially those tritium labelled compounds most widely used in life sciences research. In addition, further examples are highlighted where unexpected labelling patterns have been observed, which have led to insights into reaction mechanisms. It should be noted that the precise value recorded for any given chemical shift depends upon the condition under which the measurements have been made (pH, type of solvent, concentration, referencing, etc.). Unless otherwise stated, the data presented in this text have been derived from ^{3}H nmr spectra run at 96 MHz and with proton spin decoupling. Because tritium labelling is generally at low isotopic abundance (less than 0.1 per cent.), the ^{1}H nmr spectra cannot suffice for ^{3}H detection. This is perhaps an obvious point and is analogous to the fact that ^{13}C at its natural isotopic abundance (approximately 1 per cent.) does not usually intrude in ^{1}H nmr spectra of organic compounds. Another point which should not be overlooked is that most ^{3}H nmr spectra will be superpositions of the spectra of differently monolabelled species and not the spectrum of a single molecular species, as is common in ^{1}H nmr spectroscopy. This has considerable significance not only for the correct interpretation of ^{3}H nmr spectra where splitting patterns arise from multiply labelled compounds but also for the correct interpretation of ^{1}H-coupled ^{3}H nmr spectra.

Because of the difficulty of exactly reproducing preparative experimental conditions, particularly where heterogenous metal catalysts are involved, some degree of variation in the distribution pattern of the tritium atoms within a compound might be expected. However, in a number of cases discussed in this text, for example [G-^{3}H]phenylalanine, [3,4-^{3}H]proline, [G-^{3}H]benz[a]anthracene and [G-^{3}H]benzo[a]pyrene, the degree of batch-to-batch variation in the patterns of labelling is small, except where the quality of the starting material is variable, as for [6-^{3}H]benzo[a]pyrene. In order to assist the reader in relating observed tritium nmr spectra with the methods used for the preparation of the tritiated compound, the following code letters are used in the tables in this section:

A. Reaction of intermediate with [^{3}H]methyl iodide.
B. Reaction of intermediate with sodium borotritide.
C. Catalytic tritiation of unsaturated intermediate with tritium gas.
D. Catalytic tritiodehalogenation with tritium gas.
E. Catalysed exchange in solution with tritium gas.
F. Wilzbach exchange using tritium gas.
G. Metal catalysed exchange with tritiated water.
H. Acid catalysed exchange with tritiated water.
I. Base catalysed exchange with tritiated water.
J. Exchange with tritiated water using other catalysts.
K. Other methods of preparation.

(i) Tritium labelled amino acids (and some peptides)

A substantial number of tritiated amino acids have been analysed by ^{3}H nmr spectroscopy including all the naturally occurring amino acids [4, 25, 26, 70].

One of the first to be studied was [G-^{3}H]phenylalanine [25, 26]. This compound was prepared by platinum catalysed exchange in the presence of tritiated water at 135 °C for 18 hours and the labile tritium in the amino and carboxyl groups was subsequently removed as a step in the purification of the compound. At the level of tritium labelling achieved, with only monolabelled species present, the proton-decoupled ^{3}H nmr spectrum (Figure 12) showed a single line from each labelled position, three at low field from the phenyl ring and three at high field from the side chain. The assignments were made in part from the known ^{1}H nmr spectrum and in part from samples specifically labelled at ring positions.

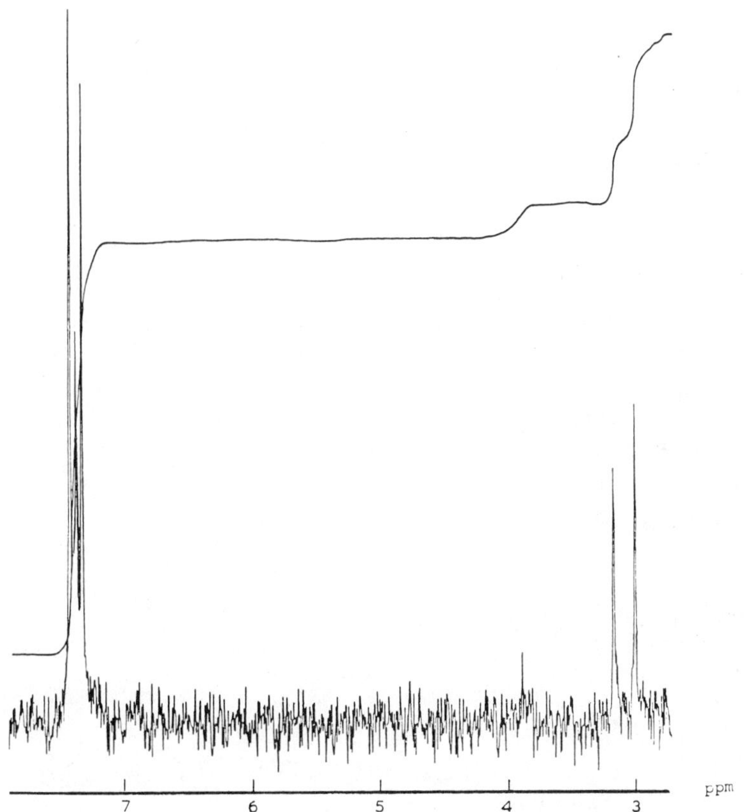

Figure 12. ^{3}H Nmr spectrum (^{1}H-decoupled) of [G-^{3}H]phenylalanine in d_6-DMSO. The spectrum in D_2O was virtually identical. (Reproduced by permission of John Wiley & Sons Ltd.)

Four separate batch preparations of L-[G-^{3}H]phenylalanine were investigated. The distribution of tritium for each batch is shown in Table 15. The data confirm that variations in the pattern of labelling from sample to sample are small. In general about 26 per cent. of the tritium is in the side chain and the

Table 15. Chemical shifts and distributions of labelling in L-[G-^{3}H]phenylalanine in D_2O. (Reproduced by permission of John Wiley & Sons Ltd.)

Chemical shift δ (p.p.m.)	Assignment	Relative intensity (%) (different samples)			
		(a)	(b)	(c)	(d)
3.06	β-CH_2	14.4	12.7	11.3	13.0
3.22	β-CH_2	12.5	12.1	7.4	11.4
3.90	α-CH	2.3	1.8	3.6	1.1
7.34	*o*-H	26.6	27.4	77.7	74.5
7.39	*p*-H	14.9	17.5		
7.43	*m*-H	29.3	28.5		

remainder in the ring. After making due allowance for there being two *meta* and two *ortho* positions, it can be seen that the ring is almost uniformly labelled with tritium.

This example demonstrates the power of the ^{3}H nmr spectroscopic method. The ability to decouple the ^{1}H interactions enables not only the chemical shifts of the three ring hydrogens (tritons) to be read off the spectrum directly but also gives the chemical shifts of the two non-equivalent β-methylene hydrogens (tritons). The complete analysis took little more than the spectral acquisition time of 8.5 hours for a 54 mCi sample; with quadrature detection, this time can now be halved. By contrast, the pattern of labelling by chemical degradation [26] is demanding in terms of time (1 man-month) and skill. In order to obtain all the necessary information the degradation scheme outlined in Table 16 was under-

Table 16. Chemical degradation scheme for [G-^{3}H]phenylalanine

DL-[G-^{3}H]Phenylalanine

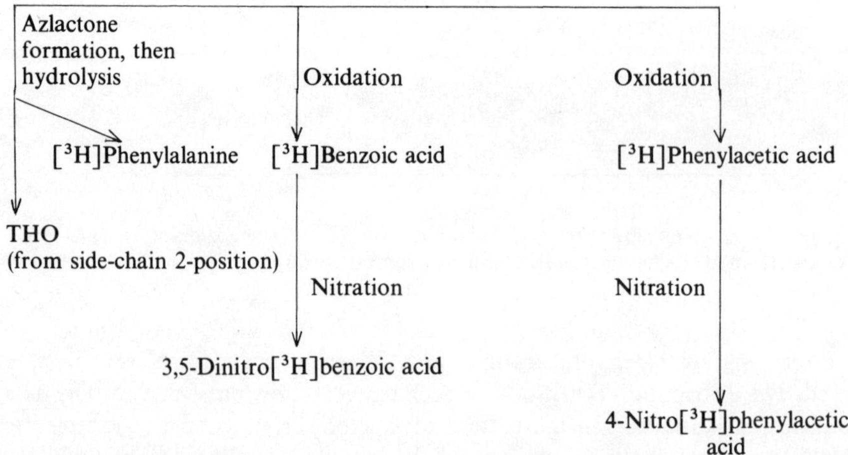

taken. The amount of tritium in the ring was determined by oxidation to [^{3}H]benzoic acid and to [^{3}H]phenylacetic acid followed by nitration, respectively, to 3,5-dinitro[^{3}H]benzoic acid and to 4-nitro[^{3}H]phenylacetic acid. Care was taken not to cause exchange of tritium from the benzylic methylene position in the latter reaction. Tritium in the ring 2- and 6-positions was calculated by difference, whereas that in the side-chain 2-position was determined by labilization of tritium during azlactone formation. Tritium in the side-chain 3-position was calculated by difference.

The overall results, namely ca. 30 per cent. of the tritium in the side chain and 70 per cent. in the ring, obtained using two different samples of DL-[G-^{3}H]phenylalanine, are in excellent agreement with those obtained much more readily by ^{3}H nmr spectroscopy, and are summarized in Table 17. In contrast, Herbert and Nicolson [71] found all the radioactivity in the ring, using a different degradation procedure. Non-specific tritium exchange during their process is the likely explanation for the different conclusions. There is a discrepancy between chemical and nmr results for tritium in the individual ring positions (see Table 17), probably to be explained by exchange processes during the nitration of the phenylacetic acid. In any case the ^{3}H nmr spectrum gives such a direct demonstration of the actual detailed distribution (Figure 12) that there is little room left for argument. In order to confirm the ring assignments, the ^{3}H nmr spectrum of specifically (and uniformly) labelled L-[2,4,6-^{3}H]phenylalanine was examined (Figure 13). The compound, prepared by catalysed halogen–tritium replacement of the trihalogeno precursor with a mixture of tritium and hydrogen, showed only two singlets, at $\delta = 7.35$ and 7.39, with intensities in the ratio of 2 : 1. The higher field signal arises from tritium at the 2,6-positions and the lower field signal from the 4-position: this last was confirmed by the results of the ^{3}H nmr (^{1}H-decoupled) spectrum of L-[4-^{3}H]phenylalanine which consisted of a single line at $\delta = 7.39$.

The proton-decoupled ^{3}H nmr spectrum of *p*-fluoro[G-^{3}H]phenylalanine (Figure 14) resembled that of [G-^{3}H]phenylalanine in the aliphatic side-chain

Table 17. Distribution of the tritium label in DL-[G-^{3}H]phenylalanine [25].

	Postition	DL-[G-^{3}H]Phenylalanine (sample 1) Degradation (%)	DL-[G-^{3}H]Phenylalanine (sample 1) ^{3}H nmr (%)	DL-[G-^{3}H]Phenylalanine (sample 2) Degradation (%)	DL-[G-^{3}H]Phenylalanine (sample 2) ^{3}H nmr (%)
CO_2H–CH(NH_2)–	2	4	2	4	4
CH_2	3	26*	25	24*	18.5
Ring	*o*	9	27		
	m	28 } 70	28.5 } 73	} 72	} 77.5
	p	33	17.5		

* By difference.

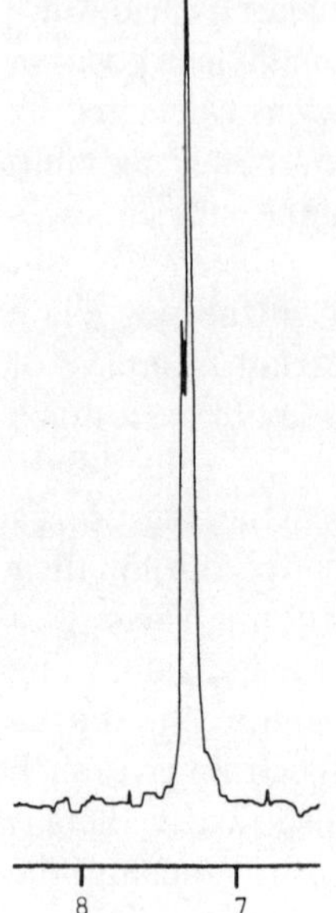

Figure 13. ^3H Nmr spectrum (^1H-decoupled) of L-[2,4,6-^3H]phenylalanine in D_2O. (Reproduced by permission of The Royal Society of Chemistry.)

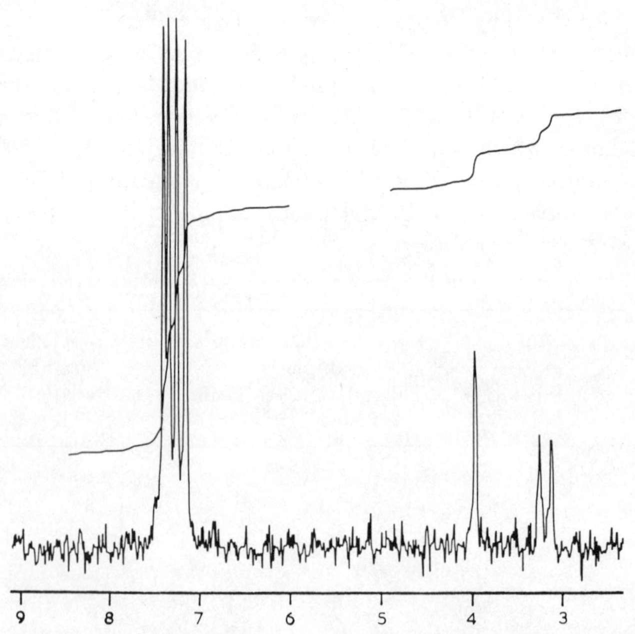

Figure 14. ^3H Nmr spectrum (^1H-decoupled) of *p*-fluoro-DL-[G-^3H]phenylalanine in D_2O. (Reproduced by permission of John Wiley & Sons Ltd.)

region but showed four lines of roughly equal intensity in the aromatic region. Since on average $J\,(o\text{-}^1HF)$ is approximately 8.5 Hz, being numerically larger than for *m*- or *p*-coupling [72], and accepting that $J(^3H,X) = 1.06664\,J(^1H,X)$ (see page 9), it follows that the higher field pair of lines, centred at $\delta = 7.17$ with $J = 9.7$ Hz, arises from the 3,5-tritons. The other doublet at $\delta = 7.33$ with $J = 5.3$ Hz comes from the 2,6-tritons: $J\,(m\text{-}^1H, F)$ on average is 5.5 Hz [72]. There is no other realistic way of pairing the observed few lines so these assignments are confirmed.

When a tritium labelled compound is prepared by the catalytic reduction of a double bond with tritium gas, there is the possibility that mono- as well as ditritiated products will be produced, either by exchange reactions occurring with the solvent or through the presence of hydrogen in the tritium gas. The ^3H nmr spectrum will then show two groups of three lines because the observed spectrum is the superposition of the spectra from doubly and the two singly tritiated species. The doubly labelled species results in the usual pair of doublet signals but inside each doublet, usually off-centre because of the changed isotope effect, is a singlet from the respective singly labelled species [40]. Such examples of superimposed spectra are numerous in ^3H nmr spectroscopy, as already mentioned, because of the relatively low isotopic abundance generally achieved for many tritium compounds.

The proton decoupled ^3H nmr spectrum of L-[3,4-^3H]proline (**X**), prepared by

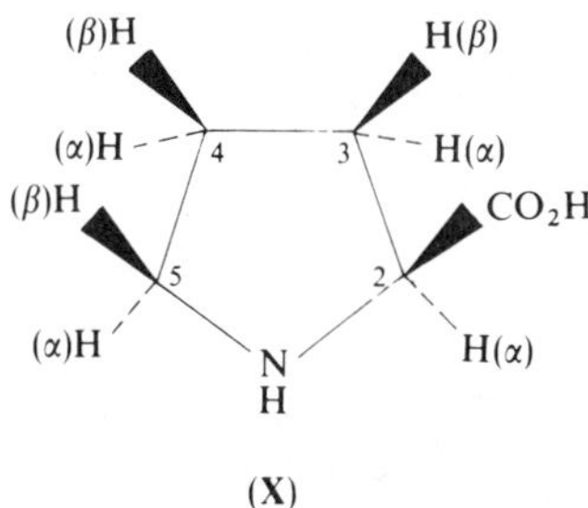

(**X**)

reduction of 3,4-dehydroproline, shows (Figure 15 and Table 18) two strong doublets at $\delta = 1.96$ and 2.32 with coupling constants of $J = 8.2$ Hz. These observations are consistent with doubly labelled *cis*-[3α, 4α-^3H]proline [25]. In addition, there are single lines at about each origin position due to the presence of singly labelled [3-^3H]- and [4-^3H]proline species. There are also two weak lines associated with tritium in the 5-positions (α and β). In the precursor, the 5-position is an allylic methylene and so subject to exchange: hence the 5-tritium is presumably introduced at that stage. Table 18 lists the quantitative results.

4-Amino[2,3-^3H]butyric acid exhibits a similar type of ^3H nmr spectrum to L-[3,4-^3H]proline, as shown in Figure 16. The aminobutyric acid was prepared from the corresponding alkene by catalytic addition of tritium gas. The spectrum shows an AB quartet ($J = 7.8$ Hz) from the mutually coupled 3- and 2-tritons, with chemical shifts of $\delta = 1.84$ and 2.24, respectively. Also visible are singlets

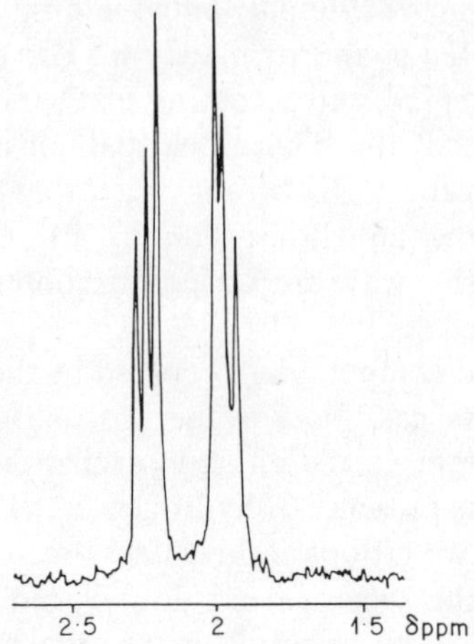

Figure 15. ^{1}H-Decoupled ^{3}H nmr spectrum of L-[3,4-^{3}H]proline in D_2O. (Reproduced by permission of Amersham International plc)

Table 18. Chemical shifts and distribution of label in L-[3,4,-^{3}H]proline in D_2O. (Reproduced by permission of John Wiley & Sons Ltd.)

Chemical shift δ (p.p.m.)	Assignment	Relative intensity (%)
1.96	4α	48.6
2.32	3α	45.7
3.24	5α	4.3
3.32	5β	
2.52	Unknown impurity	1.4

at $\delta = 1.85$ and at 2.245 from singly labelled molecules. The last two lines are shifted from the previous origin positions by the secondary isotope effect, which is therefore quite small and of the order of -0.01 to -0.005 p.p.m. (i.e. to higher field, for each additional vicinal triton). A careful analysis of the ^{3}H nmr (^{1}H-decoupled) spectrum of L-[2,3-^{3}H]proline, prepared from the 2,3-dehydrocompound by catalytic addition of tritium, revealed the presence of five isotopically distinct molecular species, three being monotritiated and two being ditritiated [70]. An even higher number of monotritiated species were produced when L-[G-^{3}H]proline was prepared by a heterogeneous catalysed exchange procedure [70].

L-Proline is also one example of a compound which has been successfully labelled by the Wilzbach technique [23]. The proton-decoupled ^{3}H nmr spectrum of L-[G-^{3}H]proline is shown in Figure 17 and the distribution of tritium derived from the spectrum is shown in Table 19. This method of labelling,

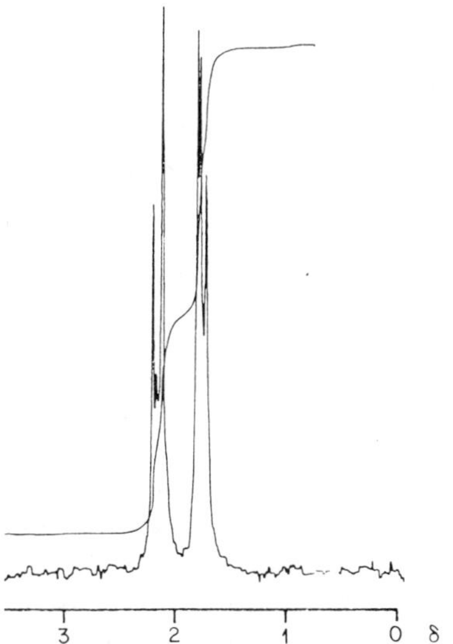

Figure 16. ^{1}H-Decoupled ^{3}H nmr spectrum of 4-amino[2,3-^{3}H]butyric acid in D_2O. (Reproduced by permission of The Royal Society of Chemistry.)

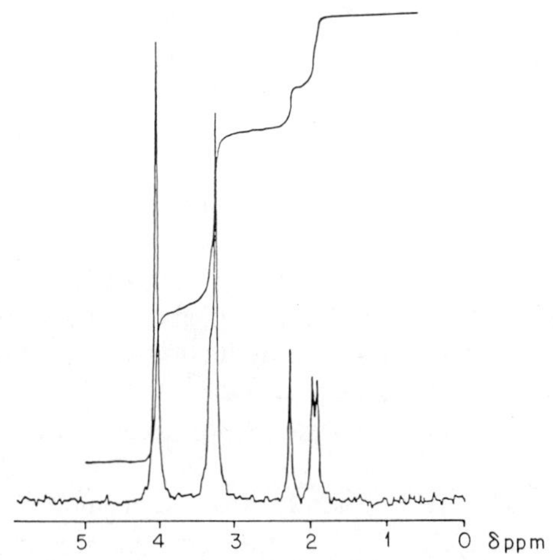

Figure 17. ^{1}H-Decoupled ^{3}H nmr spectrum of L-[G-^{3}H]proline in D_2O. (Reproduced by permission of Amersham International plc)

which favours the two positions adjacent to the nitrogen atom, is surprisingly specific. It is again stressed that chemical shift values depend upon the experimental conditions employed in their measurement (e.g. pH, solvent, concentration and temperature) so that direct comparisons with results of other investigators may require knowledge of the effects of different known experimental conditions (see pages 4 to 5).

Table 19. Chemical shifts and distribution of labelling in L-[G-^{3}H]proline in D_2O.

Chemical shift δ (p.p.m.)	Assignment	Relative intensity (%)
4.06	2	34.9
2.31	3α	9.9
2.01	3β	7.1
1.94	4	8.7
3.28	5α (+β)	39.3

The synthesis of tritiated valine provides a further interesting example where ^{3}H nmr spectroscopy has revealed quite unexpected distributions of tritium. The unsaturated intermediate *N*-acetyl-2,3-dehydrovaline (**XI**) was reduced with tritium gas using a palladium on charcoal catalyst in acetic acid [23]. After resolution with acylase-I the resulting tritiated L-valine was analysed by ^{3}H nmr spectroscopy.

$$(CH_3)_2C{=}C(COOH){-}NHCOCH_3$$

(XI)

Instead of the expected labelling in the 2- and 3-positions, a major proportion of the tritium was found in the 4-methyl groups and the distribution of tritium was shown to be 8, 28 and 64 per cent. in the 2-, 3- and 4-positions, respectively. Intramolecular hydrogen (tritium) rearrangements have been reported previously to occur during the catalytic reduction of $\alpha\beta$-unsaturated carbonyl compounds using platinum or palladium catalysts [73], and it appears that this stage rather than the enzymatic one was responsible for the unexpected labelling pattern.

Recently [23], it has been demonstrated that use of a homogeneous catalyst can give specific labelling at the 2- and 3-positions of valine via hydrogenation of **XI**. The specific radioactivity was near theoretical and the two positions were equally labelled.

The ^{3}H nmr spectroscopic examination of DL-[2, 3, 4, 5-^{3}H]ornithine (**XII**)

$$H_2N.CH_2.CH_2.CH_2.CH(NH_2).CO_2H$$

(XII)

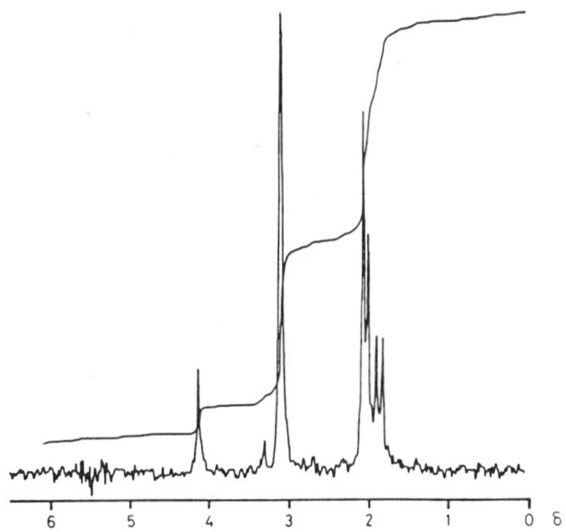

Figure 18. ^1H-Decoupled ^3H nmr spectrum of DL-[^3H]ornithine in D_2O. (Reproduced by permission of D. Reidel Publishing Company.)

provided an interesting example of superimposed spectra. Prepared via catalytic reduction of ethyl 2-acetamido-2,3-dehydro-4-cyanobutyrate, the compound (**XII**) gave a ^3H nmr spectrum (^1H-decoupled), shown in Figure 18, which exhibited four main signals, two of which comprised a pair of lines. This was not an indication of double labelling because the spacings were unequal and the intensities were incorrect for coupled doublets. The line at $\delta = 4.04$ is from the 2-position, that at $\delta = 3.03$ from the 5-position, and the lines at $\delta = 2.02$ and 1.96 evidently arise from the diastereotopic tritons in the 3-position (the two 3-tritons are non-equivalent by virtue of the 2-chirality). The pair of lines at $\delta = 1.85$ and 1.77 analogously arise from the 4-position. There is a weak line at $\delta = 3.23$ attributable to an impurity (2 per cent. of the total tritium). The whole spectrum is thus the superposition of the individual spectra of the seven species present, namely [2-^3H], [3-^3H], [3′-^3H], [4-^3H], [4′-^3H] and [5-^3H]ornithine, together with a trace [^3H]impurity.

A summary of the chemical shifts and distribution of labelling for a number of tritiated amino acids and peptides is given in Table 20.

(ii) Tritium labelled carbohydrates

The carbohydrates, or 'sugars', are amongst the most well studied classes of organic compounds and frequently used in labelled form as tracers in biochemical and biomedical research. Proven labelling is essential for many studies using tritiated carbohydrates, especially when these compounds are used as tracers for hydrogen atoms [1]. It is therefore perhaps fortunate that the methods used for tritiating carbohydrates, which generally involve the reduction of appropriate oxo-derivatives with tritiated metal hydrides, give highly specific labelling [4, 78].

Table 20. Chemical shifts and distribution of labelling for some tritium labelled amino acids in D_2O (unbuffered).

Compound	Chemical shifts δ (p.p.m.)	Assignment	Distribution of tritium (%)	Method of preparation	Reference
$HO_2C.\overset{3}{C}H_2.\overset{2}{C}H(NHCOCH_3).CO_2H$ *N*-Acetyl-DL-[2,3-^{3}H]aspartic acid	4.35 2.65 2.46	2 3 3	33 35 32	C	[23]
$(\overset{4}{C}H_3)_2\overset{3}{C}H\overset{2}{C}H(NHCOCH_3).CO_2H$ *N*-Acetyl-DL-[3,4-^{3}H]valine (in d_6-DMSO)	4.04 1.97 0.83	2 3 4	5 29 66	C	[74]
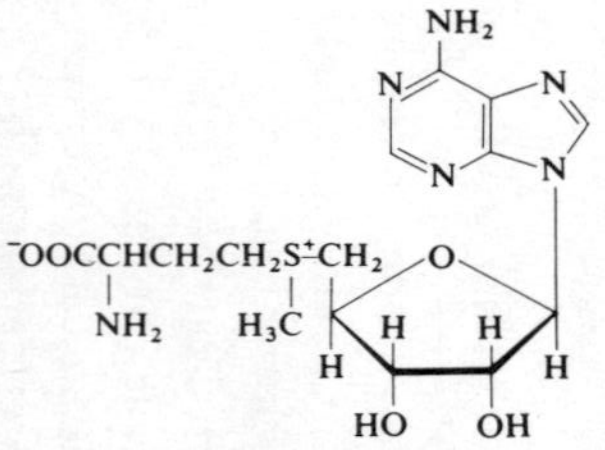 $^-OOCCH(NH_2)CH_2CH_2S^+(CH_3)CH_2$-adenosine (structure) *S*-Adenosyl-L-[*methyl*-^{3}H]methionine	2.97	Methyl	100	A	[23]

$\overset{3}{\mathrm{C}}\mathrm{H_3}\overset{2}{\mathrm{C}}\mathrm{H(NH_2).CO_2H}$ L-[2,3-^{3}H]Alanine	3.55 1.42	2 3	9.2 90.8	C	[23]
$\mathrm{H_2N.}\overset{3}{\mathrm{C}}\mathrm{H_2.}\overset{2}{\mathrm{C}}\mathrm{H_2.CO_2H}$	2.51 3.14	2 3	49.1 50.9	C	[23]
β-[2,3-^{3}H]Alanine	3.14	3	100	B	[23]
$\mathrm{H_2N.CH_2}\overset{3}{\mathrm{C}}\mathrm{H_2}\overset{2}{\mathrm{C}}\mathrm{H_2CO_2H}$ 4-Amino-*n*-[2,3-^{3}H]butyric acid	2.24 1.85	2 3	52 48	C	[23]
$\mathrm{H_2N.C(=NH)NH}\overset{5}{\mathrm{C}}\mathrm{H_2}\overset{4}{\mathrm{C}}\mathrm{H_2}\overset{3}{\mathrm{C}}\mathrm{H_2}\overset{2}{\mathrm{C}}\mathrm{H(NH_2).CO_2H}$ L-[2,3,4,5-^{3}H]Arginine mono-HCl	3.66 1.75–1.81 1.62–1.66 3.18	2 3 4 5	16.8 40.6 18.8 23.8	C	[23]
L-[5(*n*)-^{3}H]Arginine mono-HCl	1.65 3.23	4 5	9 91	C	[23]
$\mathrm{H_2NCO}\overset{3}{\mathrm{C}}\mathrm{H_2}\overset{2}{\mathrm{C}}\mathrm{H(NH_2).CO_2H}$ L-[2,3-^{3}H]Asparagine	3.86 2.82	2 3	21.5 78.5	C	[23]

Table 20. *(contd.)*

Compound	Chemical shift δ (p.p.m.)	Assignment	Distribution of tritium (%)	Method of preparation	Reference
$HO_2C.\overset{3}{C}H_2\overset{2}{C}HCO_2H$ (2-NH_2)					
L-[2,3-^{3}H]Aspartic acid	3.86	2	12.1	C	[23]
	2.75	3	35.0		
	2.69	3	52.9		
D-[2,3-^{3}H]Aspartic acid	3.88	2	32.3	C	[23]
	2.81	3	22.9		
	2.68	3	44.8		
L-[G-^{3}H]Aspartic acid	3.97	2	39.3	G	[23]
	2.94	3	25.2		
	2.89	3	35.5		
$H_2N.\overset{4}{C}H_2CH_2CHCO_2H$ (2-NH_2)					
L-2,4-Diamino[4(*n*)-^{3}H]butyric acid	2.30	3	14.6	C	[23]
	3.25	4	85.4		
F-C_6H_4-$\overset{\beta}{C}H_2\overset{\alpha}{C}H.CO_2H$ (α-NH_2; ring positions 2,3,5,6)	3.09, 3.22	β-CH_2	11	G	[25]
	3.93	α-CH	9		
	7.33	Ring-2,6	42		
	7.17	Ring-3,5	38		
p-Fluoro-DL-[G-^{3}H]phenylalanine					

$HO_2C.\overset{4}{C}H_2\overset{3}{C}H_2\overset{2}{C}H(NH_2).CO_2H$ DL-[2-^{3}H]Glutamic acid	3.73	2	100	H	[23]
L-[G-^{3}H]Glutamic acid	3.73 2.08 2.00	2 3 3	8 26 66	C	[23]
$H_2NCOCH_2CH_2CH(NH_2).CO_2H$ L-[G-^{3}H]Glutamine	3.72 2.08 2.3	2 3 4	10.8 84.4 4.8	C	[23]
$H_2N.\overset{2}{C}H_2CO_2H$ [2-^{3}H]Glycine	3.50	2	100	C	[23]
(imidazole ring, positions 2 and 5; side chain $CH_2CH(NH_2).CO_2H$) L-[2,5-^{3}H]Histidine	7.85 7.11	2 5	52.3 47.7	D	[23]
(5-hydroxyindole ring, positions 2, 4, 6, 7; side chain $CH_2CH(NH_2).CO_2H$) DL-5-Hydroxy[G-^{3}H]tryptophan	7.29 7.16 6.88 7.43	2 4 6 7	25 17.9 32.1 25	G	[23]

Table 20. *(contd.)*

Compound	Chemical shift δ (p.p.m.)	Assignment	Distribution of tritium (%)	Method of preparation	Reference
$\overset{5}{C}H_3\overset{4}{C}H_2CH(CH_3)CH(NH_2)CO_2H$ L-[4,5-^{3}H]Isoleucine	1.27 0.88	4 5	39.5 60.5	C	[23]
$(\overset{5}{C}H_3)_2\overset{4}{C}HCH_2CH(NH_2).CO_2H$ L-[4,5-^{3}H]Leucine	1.60 0.84	4 5	14 86	C	[23]
$(CH_3)_2CH.CH(NH_2)CH_2CO_2H$ β-(3RS) [2,3-^{3}H]Leucine (in CD_3OD/DCl)	2.4–2.6 3.30	2 3	31 69	C	[75]

$H_2NCH_2.CH_2CH_2CH_2\underset{\mid \atop NH_2}{C}HCO_2H$ L-[4,5-^{3}H]Lysine mono-HCl (in d_6-DMSO)	1.44 1.60 2.96	4 5 6	47.5 47.5 5.0	C	[23]
	(^{1}H-coupled spectrum)				
$CH_3SCH_2CH_2\underset{\mid \atop NH_2}{C}HCO_2H$ L-[*methyl*-^{3}H]Methionine	2.05 s 2.075 d	CT_3 CHT_2	91 9	A	[19]
$HO_2CCH_2\underset{\mid \atop NHCH_3}{C}H.CO_2H$ *N*-[^{3}H]Methyl-D-aspartic acid	2.71	Methyl	100	A	[19]
$H_2N.\overset{5}{C}H_2CH_2CH_2\underset{\mid \atop NH_2}{C}HCO_2H$	3.81 s 3.80 d	} 2	12.8		
DL-[2,3,4,5-^{3}H]Ornithine di-HCl	1.90–1.97 1.69–1.78 3.03	3 4 5	47.0 17.4 22.9	C	[23]
DL-[5(*n*)-^{3}H]Ornithine di-HCl	1.77 3.02	4 5	14 86	C	[23]
	1.74 d 3.00	4 5	7.3 92.7	B	[23]

Table 20. (*contd.*)

Compound	Chemical shift δ (p.p.m.)	Assignment	Distribution of tritium (%)	Method of preparation	Reference
(structure: phenyl ring numbered 2–6, $\overset{\beta}{CH_2}.\overset{\alpha}{CH}(NH_2).CO_2H$)					
L-[*side-chain*-^{3}H]Phenylalanine (in d_6-DMSO)	3.10 4.08	β-CH_2 α-CH	49 51	C	[23]
L-[2,3,4,5,6-^{3}H]Phenylalanine	ca. 7.4	{2, 3, 4, 5, 6}	100	G	[23]
L-[2,4,6-^{3}H]Phenylalanine	7.35 7.39	2,6 4	61 39	D	[23]
L-[2,6-^{3}H]Phenylalanine	7.34	2,6	100	D	[23]
L-[4-^{3}H]Phenylalanine	7.39	4	100	D	[25]
L-(G-^{3}H]Phenylalanine (in d_6-DMSO) (Aryl shifts are virtually the same in D_2O.)	3.06, 3.22 3.90 7.34 7.45 7.39	β-CH_2 α-CH 2, 6 3, 5 4	26 2 27 29 16	G	[25, 26]

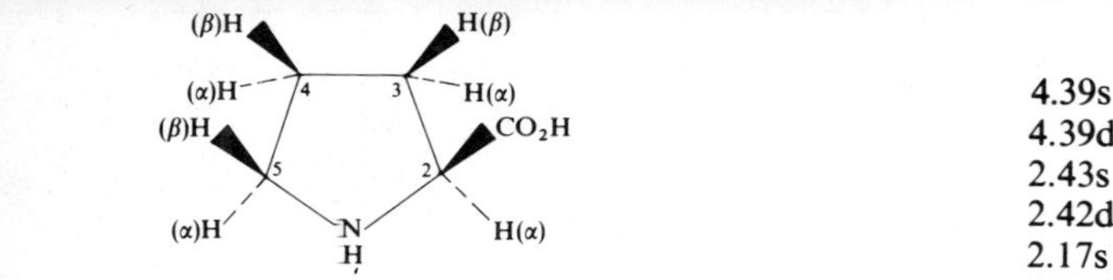

L-[2,3-^{3}H]Proline	4.39s	2α	11.9	C	[23]
	4.39d	2α coupled to 3α	13.7		
	2.43s	3α	47.2		
	2.42d	3α coupled to 2α	13.7		
	2.17s	3β	9.7		
	2.14d	3β coupled to 3α	1.9		
	2.42d	3α coupled to 3β	1.9		
L-[2,3-^{3}H]Proline (in d_6-DMSO)	4.23	2α	18.3	C	[70]
	2.23	3α	41.7		
	1.94	3β	40.0		
L-[2,3,4,5-^{3}H]Proline	4.05	2α	23.8	C	[23]
	2.28	3α	22.5		
	1.95	4α	32.2		
	3.26	5α	21.5		
L-[3,4-^{3}H]Proline	2.32	3	49	C	[25]
	1.96	4	46		
	3.30	5	4		
L-[3,4-^{3}H]Proline (in d_6-DMSO)	2.23	3α	45.4	C	[70]
	1.88	4α	49.3		
	3.14	5	5.1		
L-[4-^{3}H]Proline (in d_6-DMSO)	2.23	3α }	17	D	[70]
	1.87	4α }			
	1.88	4α	67.7		
	1.86	4β	12.8		
	3.14	5	2.5		
L-[5-^{3}H]Proline	3.29s	5α	25	B	[23]
	3.27 d	5α coupled to 5β }	50		
	3.35 d	5β coupled to 5α }			
	3.37 s	5β	25		

Table 20. *(contd.)*

Compound	Chemical shift δ (p.p.m.)	Assignment	Distribution of tritium (%)	Method of preparation	Reference
L-[G-^{3}H]Proline	4.06 2.31 2.01 1.94 3.28	2 3α 3β 4 5	34.9 9.9 7.1 8.7 39.3	F	[23]
L-[G-^{3}H]Proline (in d_6-DMSO)	4.09 2.19 1.86 1.82 3.11 3.19	2 3α 4α 4β 5α 5β	29 6 5 5 28 26	G	[70]
$HO\overset{3}{C}H_2.CH(NH_2)CO_2H$ L-[3-^{3}H]Serine	3.87	3	100	B	[23]
$H_2N.CH_2CH_2.SO_3H$ [1,2-^{3}H]Taurine	3.21 3.38	1 2	59 41	B	[76]
$CH_3.\overset{3}{C}H(OH).CH(NH_2).CO_2H$ L-[3-^{3}H]Threonine	4.30	3	100	B	[23]
DL-[G-^{3}H]Threonine	3.50 3.84	2 3	71.6 28.4	G	[23]

L-[*side-chain*-^{3}H]Tryptophan	4.00 3.43	α-CH β-CH_2	45.2 54.8	C	[23]
L-[5-^{3}H]Tryptophan	7.23	5	100	D	[25]
L-[G-^{3}H]Tryptophan (in d_6-DMSO)	2.29–3.48 6.65 7.07, 7.16 7.30, 7.44	Side chain 2 Ring	8 7 85	G	[25]
L-[*side-chain*-^{3}H]Tyrosine	3.83 2.97 3.14	α-CH β-CH_2	48 15.5 36.5	C	[23]
L-[2,3,5,6-^{3}H]Tyrosine	7.20 6.90	2, 6 3, 5	53 46	D	[23]
L-[2,6-^{3}H]Tyrosine	7.21	2, 6	100	D	[25]
L-[3,5-^{3}H]Tyrosine	6.90	3, 5	100	D	[25]

Table 20. (*contd.*) including some peptides

Compound	Chemical shift δ (p.p.m.)	Assignment	Distribution of tritium (%)	Method of preparation	Reference
$\overset{4}{C}H_3 \diagdown \overset{3}{C}H.\overset{2}{C}HCO_2H$ / $CH_3 \diagup$ with NH_2 on C-2					
DL-[2,3-^{3}H]Valine	3.555d 2.21d	2 3	50 50	C	[23]
DL-[2,3,4-^{3}H]Valine	3.54 2.20 0.95	2 3 4	21.3 30.3 48.3	C	[23]
L-[3,4-^{3}H]Valine	3.56 2.23 0.97	2 3 4	8 28 64	C	[23]
N-Acetyl-*R*-[*side-chain*-^{3}H]phenylalanyl-*S*-phenylalanine methyl ester (in d_6-DMSO)	4.40 2.75	α-CH β-CH_2	46 54	C	[77]
N-Acetyl-*S*-[*side-chain*-^{3}H]phenylalanyl-*S*-phenylalanine methyl ester (in d_6-DMSO)	4.30 2.80	α-CH β-CH_2	46 54	C	[77]
Glycyl-glycyl-*S*-[4-^{3}H]phenylalanyl-*S*-methionyl-*S*-threonyl-*S*-seryl-*S*-glutamyl-*S*-lysyl-*S*-seryl-*S*-glutaminyl-*S*-threonyl-*S*-prolyl-*S*-leucyl-*S*-valyl-*S*-threonyl-*S*-leucine	7.38	4-Phenylalanyl	100	D	[59]

S-Methionyl(sulphone)-*S*-glutamyl-*S*-[2-^{3}H]histidinyl-*S*-[4-^{3}H]phenylalanyl-*R*-lysyl-*S*-phenylalanine	8.58 7.38	2-Histidinyl 4-Phenylalanyl	100	D, E	[59]
S-Threonyl-*S*-seryl-*S*-glutamyl-*S*-[β,γ-^{3}H]lysyl-*S*-seryl-*S*-glutaminyl-*S*-threonyl-*S*-prolyl-*S*-leucyl-*S*-valyl-*S*-threonyl-*S*-leucine	1.41 1.65	γ- } Lysyl β- } Lysyl	50 50	C	[59]
[3,5-^{3}H]Tyrosyl-D-alanyl-glycyl-*N*-methylphenylalanylglyol	6.93	3, 5	100	D	[23]

The most important and widely used compound in this class is D-glucose. Labelled specifically, tritiated glucose has proved essential for investigating biochemical processes such as glycolysis and glucose synthesis in the liver [79, 80]. Table 21 summarizes the methods for introducing tritium into the six possible positions of the glucose molecule. As with many simple monosaccharides, D-glucose in aqueous solution is an equilibrium mixture of 1α- and 1β-anomeric forms. Thus for any monotritiated glucose there will be separate tritium signals arising from the two anomeric compounds present. A typical spectrum, from D-[5-^{3}H]glucose is shown in Figure 19.

Table 21. Preparation of specifically tritiated D-glucopyranose. (Reproduced by permission of John Wiley & Sons Ltd.)

Position tritiated	Method	Reference
1	Exchange with T_2/PdO–$BaSO_4$.	[57]
2	Reduction of 2-oxogluconic acid with sodium borotritide	[81]
3	Via reduction of 1:2, 5:6-di-*O*-isopropylidene-α-D-ribohexofuran-3-ulose with sodium borotritide, then 3-epimerization of the allo-furano product by benzoate displacement of the 3-tosylate	[82]
4	Via reduction of 2,3,6-tri-*O*-benzyl-α-D-xylohexopyranoside-4-ulose with sodium borotritide	[83]
5	Via reduction of 1,2-di-*O*-isopropylidene-5-oxoglucuronic acid with sodium borotritide	[84]
6	Via reduction of 1,2-di-*O*-isopropylidene-glucuronolactone with sodium borotritide	[85]

In the ^{3}H nmr spectrum of D-[6-^{3}H]glucose there are four sharp lines (Figure 20), in contrast to the broadened signals observed without ^{1}H spin decoupling [4]. The four lines give the chemical shifts of the two 6-methylene positions in each of the 1α- and 1β-anomers. The highest and lowest field lines of the group were assigned to the non-equivalent 6-positions in the major 1β-anomer on the basis of line intensities. The non-equivalence of the methylene hydrogens has its origin, of course, in the asymmetry at the adjacent 5-position.

In the absence of steric or other directional effects on the labelling process, the nmr signal intensities give a measure of the proportions of the anomers in solution. Furthermore, as ^{3}H chemical shifts are virtually identical with ^{1}H chemical shifts, the ^{3}H data, obtained by inspection, provide an easy check on ^{1}H chemical shifts which have been derived by computer analysis of complex ^{1}H nmr spectra, as is often the case with carbohydrates. This is not to say that the assignments can always be made from ^{3}H nmr spectra. Sometimes, the assignments can only be made following a successful analysis of the complex ^{1}H nmr spectrum. What a ^{3}H nmr spectrum acquired with ^{1}H decoupling can then do is to provide a direct and simple check of the various chemical shifts. Thus it was concluded from the ^{1}H nmr spectrum of 2-deoxyglucose at 300 MHz that the

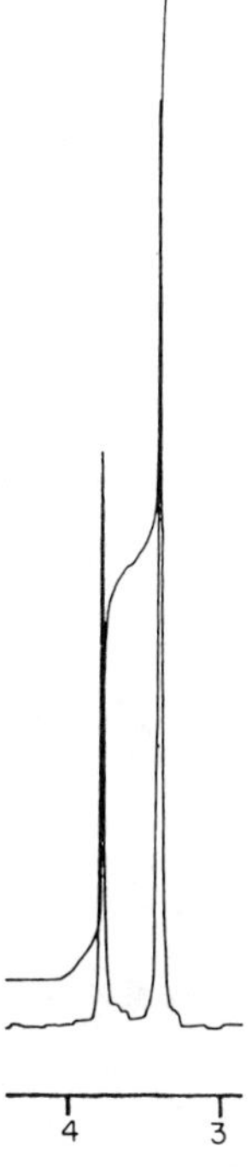

Figure 19. ^{1}H-Decoupled ^{3}H nmr spectrum of D-[5-^{3}H]glucose in D_2O, showing the signals from the 5-label in the 1α- and 1β-anomers present. (Reproduced by permission of John Wiley & Sons Ltd.)

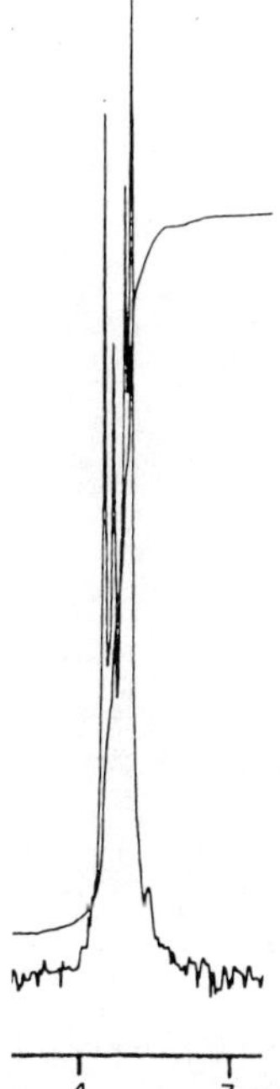

Figure 20. ^{1}H-Decoupled ^{3}H nmr spectrum of D-[6-^{3}H]glucose in D_2O. (Reproduced by permission of John Wiley & Sons Ltd.)

Table 22. Chemical shifts and distribution of labelling for some tritium labelled carbohydrates in D_2O.

Compound	Chemical shift for anomers δ (p.p.m.)		Assignment	Anomer proportion (%)		Method of preparation	Reference
	1α	1β		1α	1β		
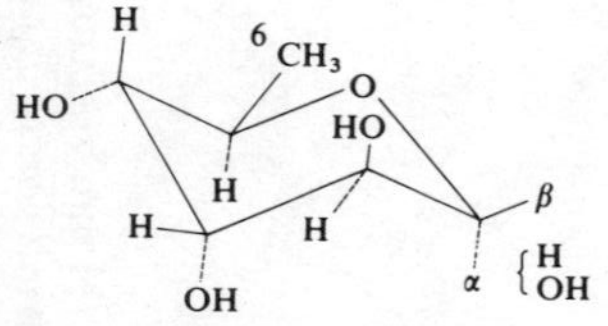 6-Deoxy-D-[6-^{3}H]altrose	1.14	1.21	6	34.8	65.2	A	[23]
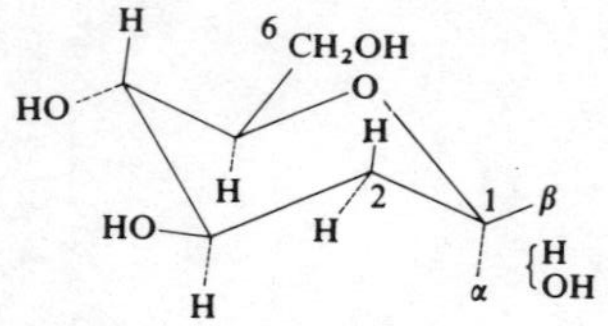 2-Deoxy-D-[1-^{3}H]glucose	5.35	4.90	1	50	50	E	[23]
2-Deoxy-D-[2,6-^{3}H]glucose	1.72	1.53	2ax.	6.1	6.1	B	[40]
	2.15	2.29	2eq.	19.8	19.8		
	3.80	3.75	6	14.7	12.7		
	3.86	3.92	6′	11.2	9.6		

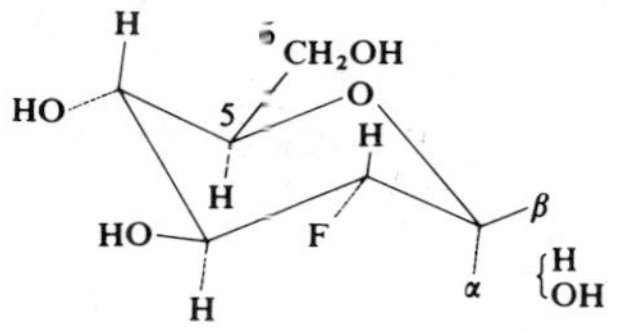 2-Fluoro-2-deoxy-D-[5,6-^{3}H]glucose	3.49 3.75 3.80	3.43 or 3.54 3.68 3.86	5 6	21.3 78.7		B	[23]
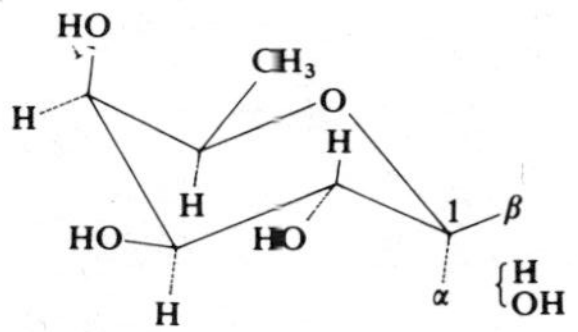 D-[1-^{3}H]Fucose	5.14	4.49	1	34	66	E	[23]
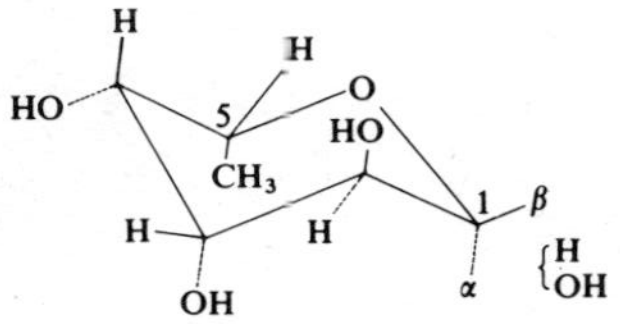 L-[1-^{3}H]Fucose	5.14	4.49	1	34.4	65.6	E	[23]

Table 22. (*contd.*)

Compound	Chemical shift for anomers δ (p.p.m.)		Assignment	Anomer proportion (%)		Method of preparation	Reference
	1α	1β		1α	1β		
L-[5,6-^{3}H]Fucose	1.21 1.19 1.15		CH_2T CHT_2 CT_3	85.2		C	[23]
	4.16 q 4.17 s	3.77 q 3.78 s	5	4.7	10.1		
L-[5,6-^{3}H]Fucose	1.24 1.21 1.18	1.21 1.19 1.14 1.14 d	6-CH_2T 6-CHT_2 6-CT_3 6-CT_3	7.5 16 15.1	14.2 25.1 18.9 3.1	C	[87]
L-[6-^{3}H]Fucose	1.12	1.16	6-CT_3	30.5	69.5	A	[23]
L-[6-^{3}H]Fucose	1.18	1.22	6	100		D	[87]
D-[1-^{3}H]Galactose	5.20	4.53	1	34.3	65.7	E	[23]
D-[6-^{3}H]Galactose	3.7		6	100		B	[23]

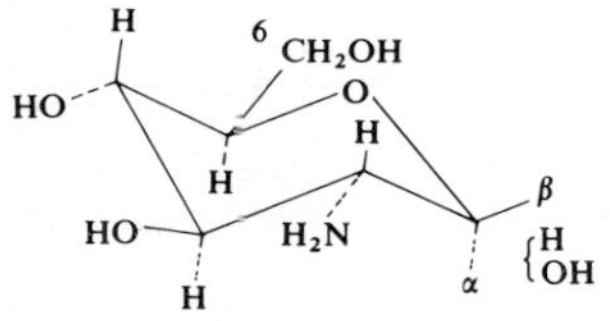

D-[1-^{3}H]Glucosamine	5.42	4.93	1	63	37	E	[23]
D-[6-^{3}H]Glucosamine	3.88 3.82 3.76 3.7		6	100		B	[23]
D-[1-^{3}H]Glucose	5.15	4.57	1	37.5	62.5	E	[78]
D-[2-^{3}H]Glucose	3.49	3.20	2	41	59	B	[78]
D-[3-^{3}H]Glucose	3.66	3.44	3	42	58	B	[78]
D-[4-^{3}H]Glucose	ca.3.40	3.36	4	100		B	[78]
D-[5-^{3}H]Glucose	3.77	3.41	5	40	60	B	[78]
D-[6-^{3}H]Glucose	3.71 3.79	3.67 3.84	6	21.5 19	32 27.5	B	[78]

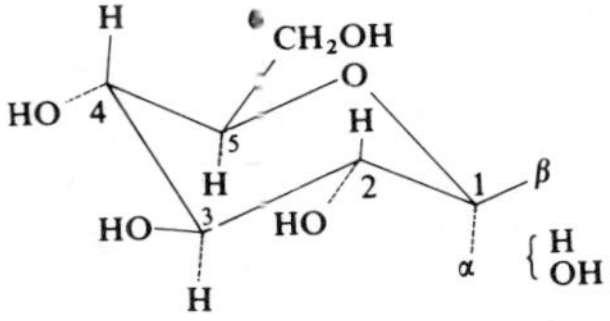

Table 22. (*contd.*)

Compound	Chemical shift for anomers δ (p.p.m.)		Assignment	Anomer proportion (%)		Method of preparation	Reference
	1α	1β		1α	1β		
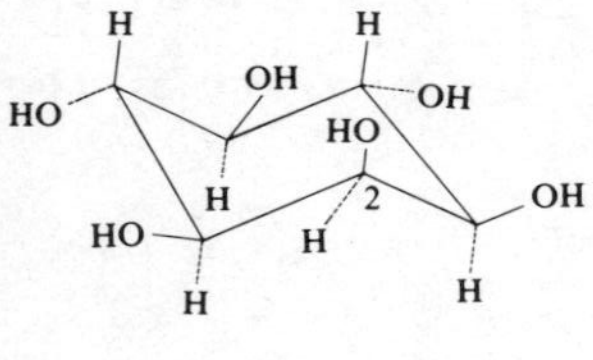 myo-[2-^{3}H]Inositol	4.03		2	100		**B**	[23]
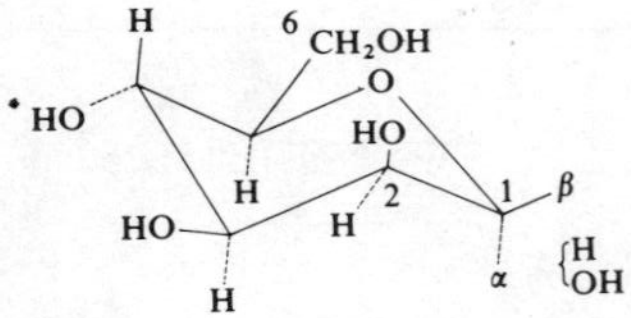 D-[1-^{3}H]Mannose	5.11	4.84	1	61	39	**E**	[23]
D-[2-^{3}H]Mannose	3.86		2	100		**B**	[23]
D-[6-^{3}H]Mannose	3.83	3.87	6	36	21	**B**	[23]
	3.73	3.70	6	32	19		

6- and 6′-chemical shifts in the 1α-anomer are identical in D_2O [86]. The ^3H nmr spectrum (with ^1H decoupling) clearly shows they are distinct, although close [54].

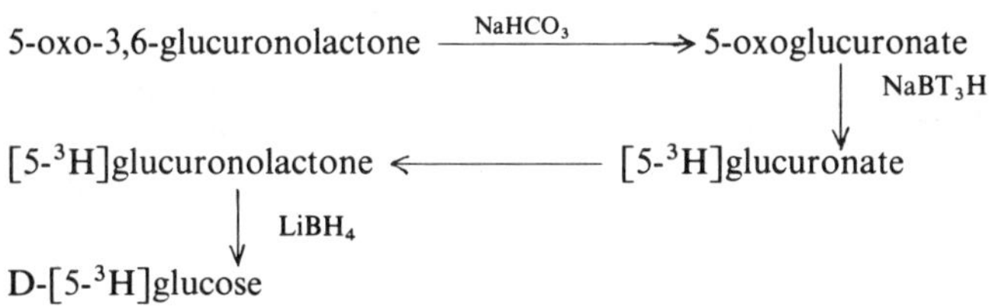

Figure 21

Although the use of tritiated metal hydrides for the reduction of oxo-derivatives usually leads to highly specific labelling in carbohydrates, the importance of using the correct and highly purified intermediate is well illustrated for the synthesis of D-[5-^3H]glucose. In this synthesis, 5-oxo-3,6-glucuronolactone is hydrolysed with sodium bicarbonate to 5-oxoglucuronate, which is then reduced with sodium borotritide to give D-[5-^3H]glucose by the reaction scheme shown in Figure 21. However, the ^3H nmr spectrum (Figure 22) revealed that the product was D-[5,6-^3H]glucose. Subsequent studies showed that if the hydrolysis of the 5-oxo-3,6-glucuronolactone is incomplete, reduction with sodium borotritide yields a mixture of D-[5,6-^3H]glucose and [5-^3H]glucuronate. The further reactions then lead finally to a mixture of D-[5,6-^3H]glucose and the required D-[5-^3H]glucose. The constant necessity for ensuring the highest purity for each intermediate—in this case 5-oxoglucuronate—was thus further emphasized.

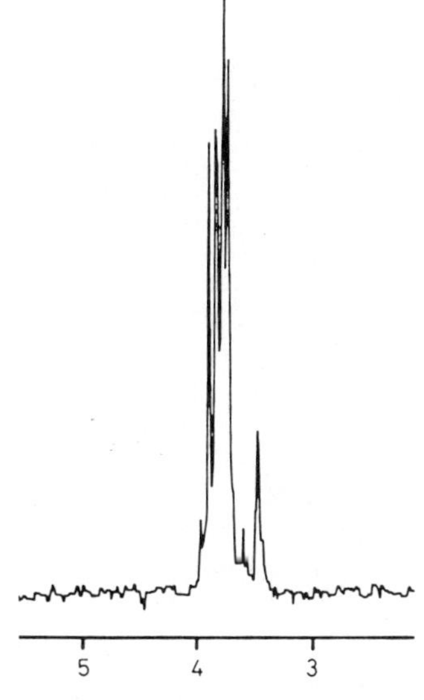

Figure 22. ^1H-Decoupled ^3H nmr spectrum of D-[5,6-^3H]glucose in D_2O.

The chemical shifts and patterns of labelling for some tritium labelled carbohydrates are shown in the Table 22.

(iii) Tritium labelled polycyclic aromatic hydrocarbons

Polycyclic aromatic hydrocarbons have been labelled with tritium by a variety of heterogeneous and homogeneous catalytic methods and their ^3H nmr spectra have provided an insight into the mechanisms of action of the catalysts involved. This aspect is discussed in detail in Chapter 3.

Polycyclic aromatic hydrocarbons labelled with tritium are widely used in studies of the mechanisms of carcinogenesis. The use of tritium compounds as

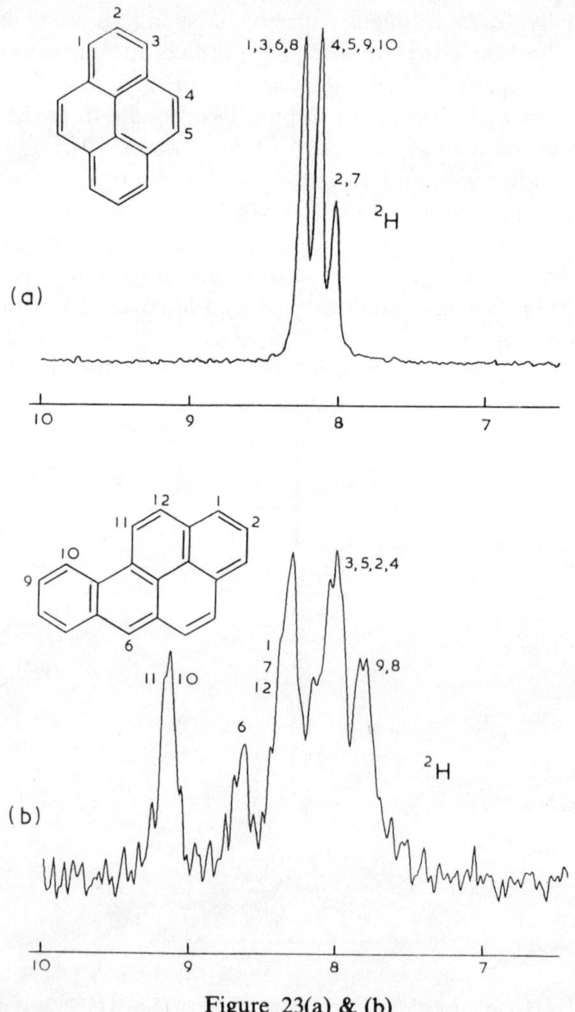

Figure 23(a) & (b)

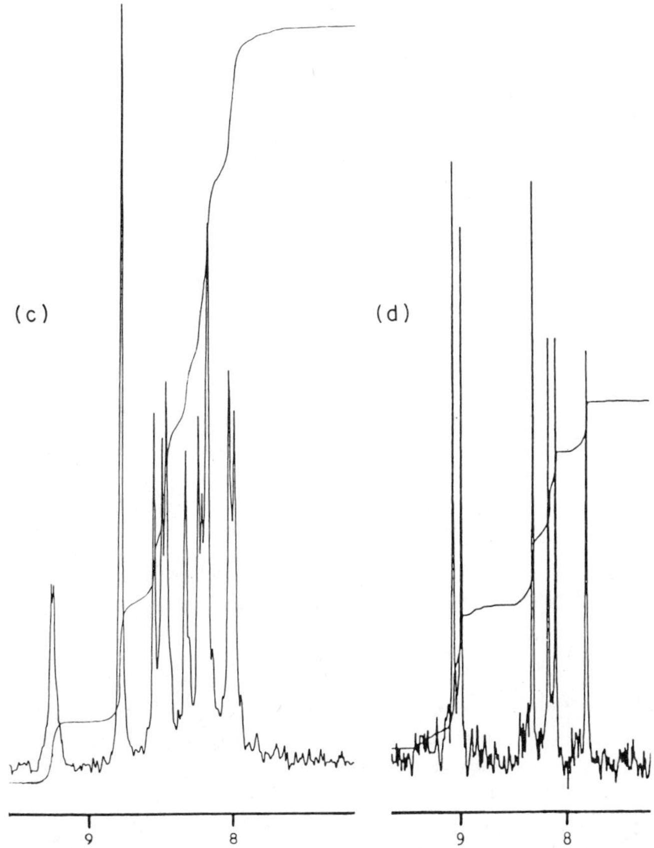

Figure 23. (a) ^{2}H Nmr spectrum (61.424 MHz; ^{1}H-decoupled) of [G-^{2}H]pyrene. All three different sites are labelled and the lines, which are 0.11 p.p.m. apart, are properly resolved, although somewhat broad as compared with the lines in ^{3}H or ^{1}H spectra [91,92]. (b) ^{2}H Nmr spectrum (61.424 MHz; ^{1}H-decoupled) of [G-^{2}H]benzo[a]pyrene. Here the lines are too close to be resolved [91,92]. (c) ^{3}H Nmr spectrum (^{1}H-decoupled) of [G-^{3}H]benzo[a]pyrene and (d) [G-^{3}H]benzo[e]pyrene in d_6-DMSO.

tracers in these studies has undoubtedly been hindered by the lack of suitable methods, such as specific degradation or substitution reactions, for determining their patterns of tritium labelling. One of the early successes of ^{3}H nmr spectroscopy was in determining the tritium distribution in a series of generally labelled polycyclic aromatic hydrocarbons [27], prepared by platinum catalysed exchange in 70 per cent. acetic acid containing tritiated water [1]. In most cases the ^{1}H nmr spectra have been fully analysed [88 to 90] so that a careful comparison of the ^{3}H with the ^{1}H data gives the assignments, i.e. the positions labelled as well as the relative amounts of tritium present in each position. Furthermore, because the chemical shifts in the aromatic region are so close together, normally all within 2 p.p.m., the alternative of using deuterium labelled

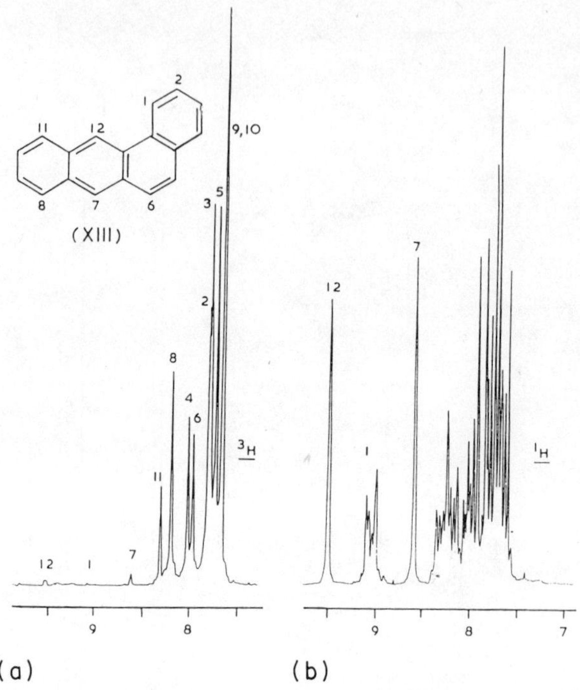

Figure 24. (a) ^{3}H Nmr spectrum (^{1}H-decoupled) of [G-^{3}H]benz[a]anthracene. The attainable resolution at 96 MHz is clearly high. (b) ^{1}H Nmr spectrum of benz[a]anthracene, a typical high resolution spectrum at 90 MHz. (Reproduced by permission of D. Reidel Publishing Company.)

compounds and ^{2}H nmr spectroscopy (Figure 23a) is by no means always applicable. Because ^{2}H nmr lines tend to be broad and particularly because the spectral dispersion is relatively poor, the close aromatic lines are not always resolved (see Figure 23b). In contrast, ^{3}H nmr lines are very narrow and well resolved, as in Figure 23c which shows the proton-decoupled ^{3}H nmr spectrum of tritiated benzo[a]pyrene, along with that of benzo[e]pyrene in Figure 23d. Whilst the ^{1}H-decoupled ^{3}H nmr spectrum (Figure 23c) gives the chemical shifts of the tritiated positions direct and the relative degree of labelling from the line intensities, the assignments can only be made by reference to the computer analysis of the complex ^{1}H nmr spectrum [90], taking into account the effects of solvent, concentration and temperature differences, if any.

The chemical shifts and distribution of labelling for a number of polycyclic aromatic hydrocarbons are given in Table 23.

The ^{3}H nmr results from the labelling of polycyclic aromatic hydrocarbons (Table 23) indicate steric hindrance effects in these catalysed preparation reactions. Thus for [G-^{3}H]benz[a]anthracene (Figure 24a), no detectable labelling occurs at the 1- and 12-positions whilst labelling at the 7-position is low, perhaps due to hindrance from the two adjacent *peri*-positions. With 7-methylbenz-

Table 23. Chemical shifts and distribution of labelling for some tritium labelled polycyclic aromatic hydrocarbons

Compound	Solvent	Chemical shift δ (p. p. m)	Assignment	Distribution of tritium (%)		Method of preparation	Reference
		7.49	1,4,5,8	44			
		8.03	2,3,6,7	38			
[G-^{3}H]Anthracene	$CDCl_3$	8.46	9,10	18		G	[13]
				(a)	(b)		
		8.95	1	0	0		
		7.78	2	22	12		
		7.76	3		12.5		
		7.97	4	9	7.5		
		7.70	5	13	15.5		
		7.92	6	11	7		
		8.55	7	2	1	G	[27]
		8.16	8	10	10		
		7.63	9,10	23.5	30		
		8.27	11	9	5		
[G-^{3}H]Benz[a]anthracene	d_6-DMSO	9.37	12	0	0		
		8.41	1	10	11		
		8.14	2	7	18		
		8.11	4	12			
		8.27	3	10	9		
		8.17	5	8	10		
		8.76	6	10	13.5		
		8.44	7	8	4.5	G	[27]
		7.90	8	9	10		
		7.94	9	12	12		
		9.26	10	3	2		
		9.28	11				
[G-^{3}H]Benzo[a]pyrene	d_6-DMSO	8.51	12	11	9		

Table 23. (*contd.*)

Compound	Solvent	Chemical shift δ (p. p. m)	Assignment	Distribution of tritium (%)		Method of preparation	Reference
		8.43	1	9.6			
		8.16	2	6.2			
		8.28	3	9.6			
		8.12	4	12.4			
		8.18	5	6.7			
		8.78	6	8.4			
		8.46	7	7.3			
		7.91	8	6.7		H	[23]
		7.95	9	12.4			
		9.28	10	4.2			
		9.31	11	4.2			
[G-^{3}H]Benzo[a]pyrene	d_6-DMSO	8.52	12	12.4			
				(a)	(b)		
		8.43	1	4.0	16.7		
		8.28	3	1.6	8.9	D	[27]
[6(n)-^{3}H]Benzo[a]pyrene	d_6-DMSO	8.78	6	94.4	74.4		
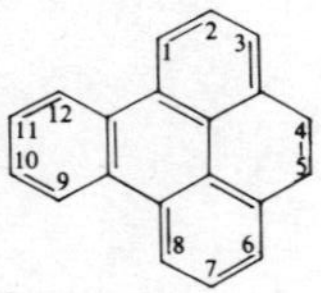		9.10	1,8	19			
		8.14	2,7	14.5			
		8.35	3,6	20			
		8.21	4,5	14.5		G	[92]
		9.03	9,12	18			
[G-^{3}H]Benzo[e]pyrene	d_6-DMSO	7.84	10,11	14			

		8.51	1	0		
		7.74	2			
		7.73	3	43		
		7.72	5			
		7.97	4	18		
		8.11	6	0		
		3.03	7-methyl	6		
		8.44	8	5		
		7.66	9	11	G	[27]
		7.67	10	16		
		8.40	11	0		
[G-^{3}H]7,12-Dimethylbenz[a]anthracene	d_6-DMSO	3.26	12-methyl	1.5		
		9.04	1	0.5		
		7.71	2	10		
		7.76	3	12		
		8.00	4	16		
		7.83	5	13		
		3.08	7-methyl	19	G	[27]
		8.28	6,11	7.5		
		8.41	8	3		
		7.66	9	10		
		7.68	10	6		
[G-^{3}H]7-Methylbenz[a]anthracene	d_6-DMSO	9.41	12	2		

Table 23. *(contd.)*

Compound	Solvent	Chemical shift δ (p. p. m)	Assignment	Distribution of tritium (%)	Method of preparation	Reference
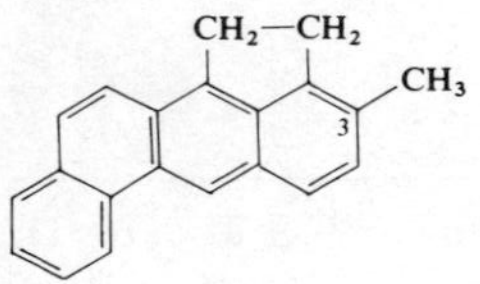		3.68	1-CH_2	12		
		3.39	2-CH_2	13		
		2.36	3-CHT_2	20		
		2.38	3-CH_2T			
		7.34	4	7.5		
		7.76	5,12	10		
		8.93	6	6	G	[27]
		8.79	7	1		
		7.63	8	8		
		7.60	9	8		
		7.82	10	7.5		
[G-^{3}H]3-Methylcholanthrene	d_6-DMSO	7.56	11	6		
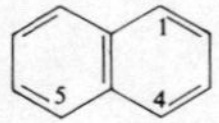		7.86	1,4,5,8	49		
(G-^{3}H]Naphthalene	$CDCl_3$	7.49	2,3,6,7	51	G	[13]

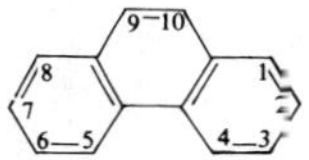 [G-^{3}H]Phenanthrene	$CDCl_3$	7.93	1,8	18	G	[13]
		7.63	2,7	21		
		7.69	3,6	23		
		8.72	4,5	19		
		7.77	9,10	19		
[G-^{3}H]Pyrene	d_6-DMSO	8.36	1,3,6,8	36	G	[13]
		8.14	2,7	29		
		8.25	4,5,9,10	35		
[G-^{3}H]Triphenylene	$CDCl_3$	8.67	1,4,5,8,9,12	59	G	[13]
		7.68	2,3,6,7,10,11	41		

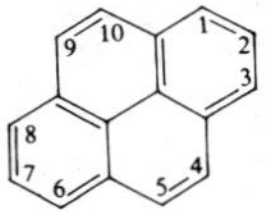

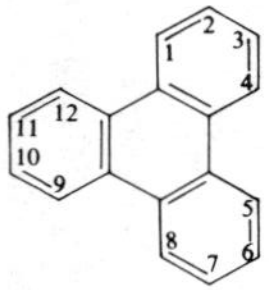

[a]anthracene, the extent of labelling is likewise very low at the 1- and 12-positions, as it is also in the 8-position. Labelling in the 6-position is probably low, as suggested by the low intensity of the superimposed 6- and 11-^{3}H signals. These results reflect the steric effects of the 7-methyl group. For 7,12-dimethylbenz[a]anthracene no detectable tritium is incorporated at the 6- or the 11-positions and very little in the 12-methyl group. The ^{1}H-decoupled ^{3}H nmr spectrum of benzo[a]pyrene (Figure 23c) shows that the incorporation of tritium label is remarkably uniform in all positions except for the sterically hindered 10- and 11-positions. The labelling procedure is consistent in that different samples give a very similar tritium distribution. However, in the case of benzo[e]pyrene (Figure 23d and Table 23) steric hindrance effects seem to be less important than for benzo[a]pyrene.

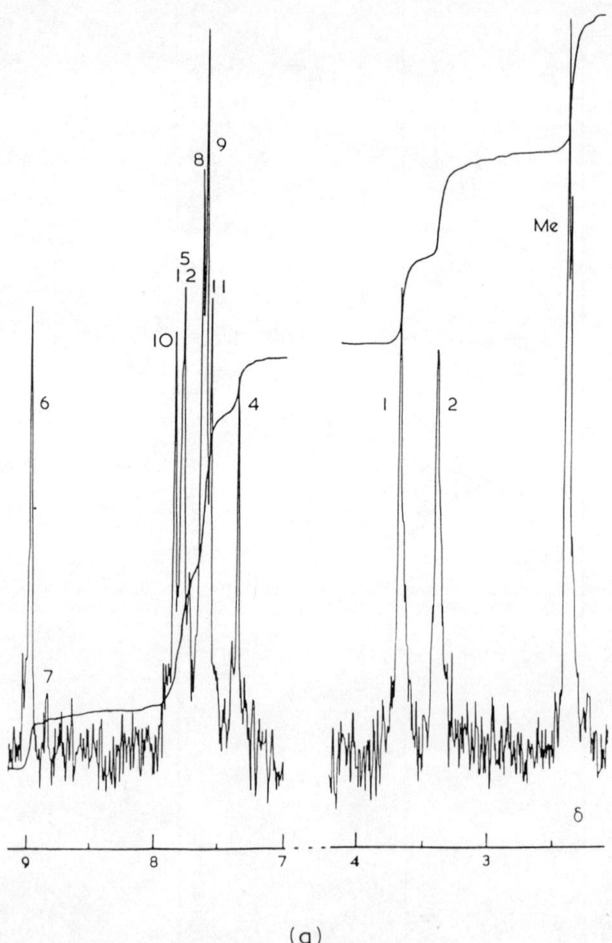

Figure 25. ^{3}H Nmr spectra of [G-^{3}H]-3-methylcholanthrene in d_6-DMSO (a) with ^{1}H-decoupling and (b) with ^{1}H-coupling. (Reproduced by permission of The Royal Society of Chemistry.)

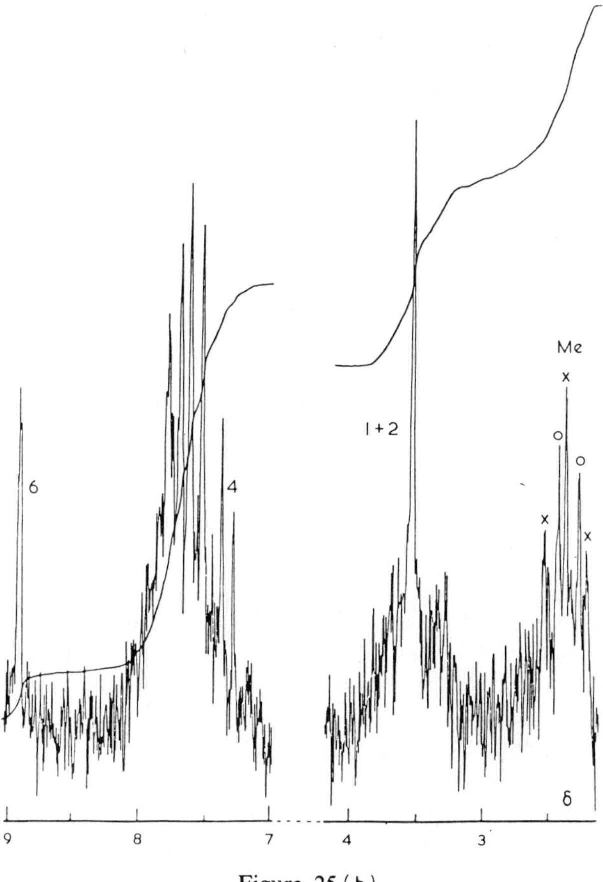

Figure 25 (b)

Interestingly the homogeneous labelling of benzo[a]pyrene in tritiated trifluoroacetic acid shows a similar pattern of tritium distribution to that obtained by the heterogeneous catalysed procedure [23]. This confirms that 'steric effects' can also occur in homogeneous catalysis, but to a lesser extent (see Table 23).

Heterogeneous labelling results for 3-methylcholanthrene (Figure 25) show that the platinum catalyst also has the ability to label both aromatic and aliphatic regions, with the aliphatic groups accounting for 45 per cent. of the tritium and the rest being thinly distributed amongst the nine aromatic positions with the 7-position containing the least amount of tritium. The 3-methyl group has two chemical shifts of $\delta = 2.36$ and 2.38 consistent with CHT_2 and CH_2T labelling, and this is confirmed in the undecoupled ^3H nmr spectrum (Figure 25b) by the appearance of a doublet ($J = 15.3$ Hz) and a triplet ($J = 15.1$ Hz), respectively.

Specific labelling in polycyclic aromatic hydrocarbons can be achieved by catalysed halogen–tritium replacement, as used, for example, in the preparation of [6-^{3}H]benzo[a]pyrene. The need for highly pure 6-bromobenzo[a]pyrene as the starting compound has already been discussed (see page 27). The chemical shifts and distribution of labelling in [6-^{3}H]benzo[a]pyrene prepared from a number of different samples of 6-bromobenzo[a]pyrene, of varied quality, are shown in Table 24.

Table 24. Chemical shifts and distribution of labelling for [6-^{3}H]benzo[a]pyrene in d_6-DMSO. (Reproduced by permission of John Wiley & Sons Ltd.)

Chemical shift δ (p.p.m.)	Assignment	Distribution of tritium in various samples (%)			
		(a)	(b)	(c)	(d)
8.78	6	90.7	81.0	74.4	94.4
8.43	1	8.2	14.8	16.7	4.0
8.28	3	1.1	4.2	8.9	1.6

(iv) Tritium labelled aromatic hydrocarbons

These compounds have been prepared mainly as a result of mechanistic studies of tritium exchange and the examples cited in Table 25 are mainly labelled generally with tritium. However, exchange conditions can be obtained which give solely side-chain labelling as in toluene [93] and ethylbenzene [14], or solely aromatic ring labelling as in heptylbenzene [94].

If specific labelling were required it could be achieved by catalytic tritiodehalogenation (ring labelling) or double bond saturation (side-chain labelling), as demonstrated later by numerous examples in Table 29.

The change of chemical shift with solvent composition has been investigated in detail for toluene and is discussed further on pages 186 to 187.

(v) Tritium labelled nucleic acid derivatives

The chemical shifts and distribution of tritium observed for some labelled purines, pyrimidines, nucleosides and nucleotides are shown in Table 26. In general these compounds are labelled with a high degree of specificity, as indeed expected from the methods of preparation.

Especial care is needed in the preparation of samples of the tritiated nucleoside triphosphates for ^{3}H nmr spectroscopy, or indeed for any other purpose, because of their extreme ease of hydrolysis. Thus the spectrum of uridine triphosphate (Figure 26) shows minor signals from an impurity and from [5,6-^{3}H]uracil species arising from hydrolytic decomposition during preparation or examination of the sample. The minor doublets (a and b) are evidently from the doubly labelled [5,6-$^3H_2$]uracil and the signal c from an unknown impurity. There is a single line inside each doublet; that at a derives from [5-^{3}H]uracil and that at b from [6-^{3}H]uracil. The major doublets centred at $\delta = 7.962$ and 5.978

Table 25. Chemical shifts and distribution of labelling for some tritium labelled aromatic hydrocarbons

Compound	Solvent	Chemical shifts δ (p.p.m.)	Assignment	Distribution of tritium (%)			Method of preparation	Reference
$[^3H]$Benzene	Neat	7.14	Ring		—		G,J	[95]
	5% in cyclohexane	7.26	Ring		—		G,J	[95]
				(a)	(b)			
n-$[^3H]$Butylbenzene	CCl_4	7.06	2,4,6	23	—		G	[93]
		7.15	3,5	15	—			
		2.46	α-CH_2	18	100			
		1.49	β-CH_2	18	—			
		1.24	γ-CH_2	13	—			
		0.85	Methyl	13	—			
sec-$[^3H]$Butylbenzene	Neat	7.07	2,6		31.8		J	[94]
		7.15	3,5		5.2			
		7.04	4		30.9			
		2.40	CH		< 1			
		1.46, 1.49	CH_2		13.4			
		1.12	β-methyl		18.6			
		0.73	γ-methyl		< 1			
				(a)	(b)	(c)		
1,4-$[^3H]$Dimethylnaphthalene	Neat	6.91	2,3	14	32	39	G,J	[93]
		7.70	5,8	4	33	33		
		7.26	6,7	39	35	28		
		2.31	Methyls	43	0	0		

Table 25. (*contd.*)

Compound	Solvent	Chemical shifts δ (p.p.m.)	Assignment	Distribution of tritium (%)	Method of preparation	Reference
1,5-[^{3}H]Dimethylnaphthalene	CCl_4	7.17	2,6	18	G	[93]
		7.24	3,7	18		
		7.72	4,8	5		
		2.56	Methyls	58		
2,6-[^{3}H]Dimethylnaphthalene	CCl_4	7.44	1,5	16	G	[93]
		7.16	3,7	15		
		7.53	4,8	15		
		2.40	Methyls	54		
[^{3}H]Ethylbenzene	Neat	1.09	CH_3	25	G	[14]
		2.46	CH_2	75		
	$CDCl_3$	(a) (b)		(a) (b)	C	[87,96]
		1.1 1.20	CH_3	80 60		
		2.6 2.72	CH_2	20 40		
n-[^{3}H]Heptylbenzene	Neat	7.04	2,6	53	J	[94]
		7.12	3,5	4.4		
		7.06	4	42.5		
		—	Alkyl	< 1		
[^{3}H]Isopropylbenzene	Neat			(a) (b)	G,J	[93]
		7.12	2,6	13 47		
		7.18	3,5	21 —		
		7.07	4	11 42		
		2.70	CH	11 —		
		1.14	Methyls	44 10		

[^{3}H]Isopropylbenzene	Neat	7.08	2,6	15			
		7.14	3,5	< 1			
		7.03	4	16.2		J	[94]
		2.55	CH	< 1			
		1.09	Methyls	69			
[^{3}H]Naphthalene	$CDCl_3$	7.86	1,4,5,8	49		G	[13]
		7.49	2,3,6,7	51			
n-[^{3}H]Propylbenzene	CCl_4	2.39	α-CH_2	100		G	[23]
				(a)	(b)		
[^{3}H]Styrene	$CDCl_3$	5.2	Vinyl *trans*	15	50	C	[87,96]
		5.8	Vinyl *cis*	56	50		
			Aromatic	29	—		
				(a)	(b)		
[^{3}H]Toluene	CCl_4	7.09	2,4,6	32	—	G	[93]
		7.17	3,5	24	—		
		2.12	Methyl	44	99		
	Neat	7.013	2,6	—		—	[97]
		7.041	3,5	—			
		7.123	4	—			
		2.117	Methyl	—			
	10% in CCl_4	7.095	2,6	—		—	[97]
		7.082	3,5	—			
		7.176	4	—			
		2.296	Methyl	—			

Table 25. *(contd.)*

Compound	Solvent	Chemical shifts δ (p.p.m.)	Assignment	Distribution of tritium (%)	Method of preparation	Reference
[³H]Toluene	10% in DMSO	7.214	2,6	—	—	[97]
		7.190	3,5	—		
		7.291	4	—		
		2.282	Methyl	—		
	Neat	7.04	2,6	—	J	[34]
		7.16	3,5	—		
		7.08	4	—		
	Neat			(a) (b)	—	[94,95]
		6.97	2,6	57.4 59		
		7.09	3,5	< 1 6.6		
		7.01	4	42.5 34.4		
	25% in CCl_4	7.07	2,6	—	–	[95]
		7.16	3,5	—		
		7.07	4	—		
	CCl_4			(a) (b)	G	[23]
		2.12	Methyl	90 20		
		7.09	Aromatic	10 80		
1,3,5-[³H]Trimethylbenzene	Neat			(a) (b)	G,J	[93]
		6.65	2,4,6	— 100		
		2.11	Methyls	100 —		
m-[³H]Xylene	CCl_4	6.90	2	6	G	[93]
		6.87	4,6	16		
		7.04	5	11		
		2.24	Methyls	66		

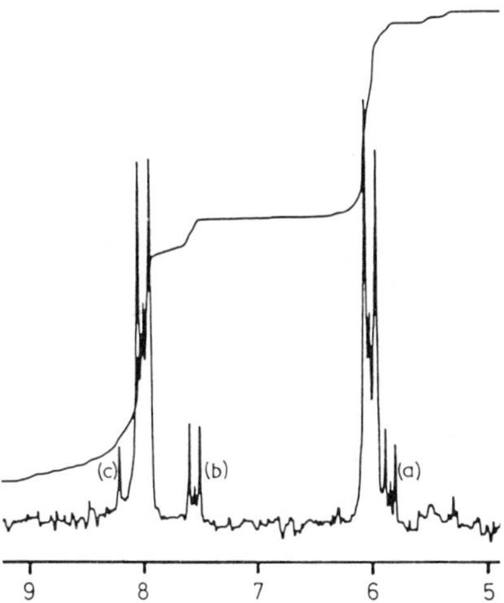

Figure 26. ^{1}H-Decoupled ^{3}H nmr spectrum of [5,6-$^3H_2$]uridine triphosphate in D_2O showing the presence of [5,6-$^3H_2$]uracil impurity and of monolabelled species. (Reproduced by permission of D. Reidel Publishing Company.)

($J = 9.2$ Hz) are from [5,6-$^3H_2$]uridine triphosphate. Inside the doublets are singlets at $\delta = 7.959$ and 5.982 which derive from [6-^{3}H]- and [5-^{3}H]-uridine triphosphate, respectively.

Detailed analysis of tritiated methyl groups in compounds is possible by ^{3}H nmr spectroscopy with the aid of resolution enhancement. A tritiated methyl group may contain the species CH_2T, CHT_2 and CT_3. Previously there has been no direct way of determining the distribution of such labelling but now with ^{3}H nmr spectroscopy this is conveniently done by integration of the tritium signals which are separated by the primary isotope effect. Because the resultant isotopic shifts are small [14], it is advantageous to enhance the resolution before attempted integration of the rather close signals. The method of Clin *et al.* [98] for resolution enhancement in Fourier transform nmr is a method of digital filtering by a sinusoidal window [99] which has proved to be most effective and is easy to apply. Although the resolution enhancement procedure necessarily distorts the line shape, there should be no relative effect on line intensities provided the relaxation time T_2* is the same for each line. This is very likely to be so for the lines from a tritiated methyl group.

The proton-decoupled ^{3}H nmr spectrum of [*methyl*-^{3}H]thymine prepared by catalysed reduction of 5-formyluracil with tritium gas showed three close lines in the $\delta = 1.81$ region of the spectrum (Figure 4a), suggesting the presence of three kinds of labelled methyl group. Resolution enhancement (Figure 4b) allowed the relative intensities of these lines at $\delta = 1.79$, 1.81 and 1.84 to be measured as 29, 54 and 17 per cent., respectively. The origin of these lines from CT_3, CHT_2 and CH_2T methyl groups was confirmed by the coupled, resolution-enhanced ^{3}H nmr spectrum (Figure 4c), the results from which are included in Table 26. The

Table 26. Chemical shifts and distribution of labelling for some tritium labelled nucleic acid derivatives in D_2O

Compound	Chemical shift δ (p.p.m.)		Assignment	Distribution of tritium (%)	Method of preparation	Reference
	pD					
	1.50	8.41				
	1.84	8.35	2	100	D	[23]
	2.57	8.21				
	6.2	8.22				
[2-^{3}H]Adenine						
[8-^{3}H]Adenine		8.17	8	100	D	[23]
[2,8-^{3}H]Adenine	7.4	8.19 8.15	2 8	50 50	D	[23]
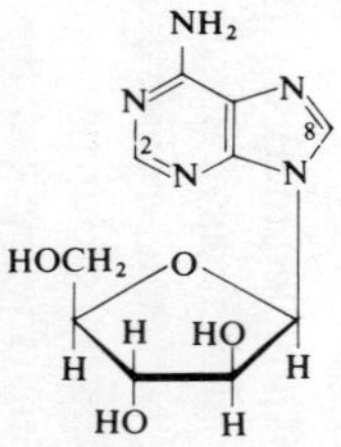 [2,8-^{3}H]Adenine-β-D-arabinoside		8.24 8.37	2 8	48 52	E	[23]

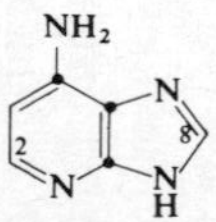

Compound	Solvent	δ (ppm)	Position	%	Method	Ref.
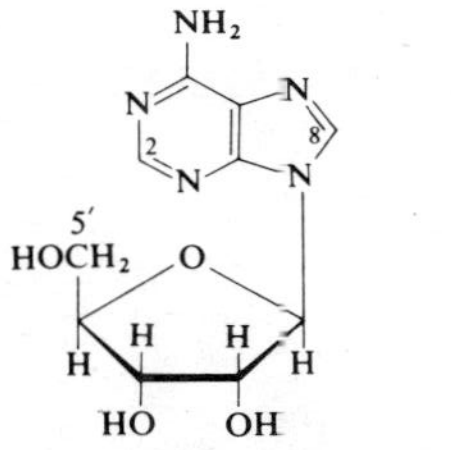 [2,5′,8-^{3}H]Adenosine	pD	3.82, 3.89	5′	38, 22	B, E	[49]
	5.0	8.26	2	4		
		8.35	8	36		
[2,8-^{3}H]Adenosine	7.3	8.23	2	22	E	[23]
		8.33	8	78		
[5′,8-^{3}H]Adenosine	7.3	3.80, 3.87	5′	68.6	B, D	[49]
		8.33	8	31.4		
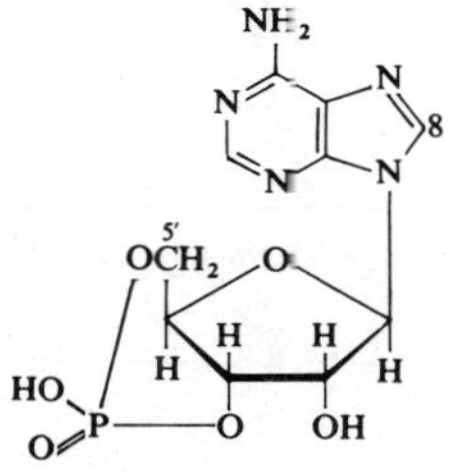 [5′,8-^{3}H]Adenosine cyclic monophosphate*		4.32, 4.59	5′	58.4	B, D	[23]
		8.27	8	41.6		

Table 26. (*contd.*)

Compound	Chemical shift δ (p.p.m.)	Assignment	Distribution of tritium (%)	Method of preparation	Reference
[8-^{3}H]Adenosine cyclic monophosphate*	8.36	8	100	E	[23]
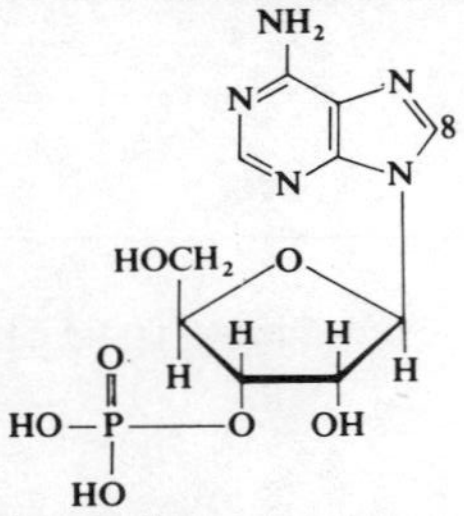 [8-^{3}H]Adenosine 3′-monophosphate*	8.20	8	100	I	[100]
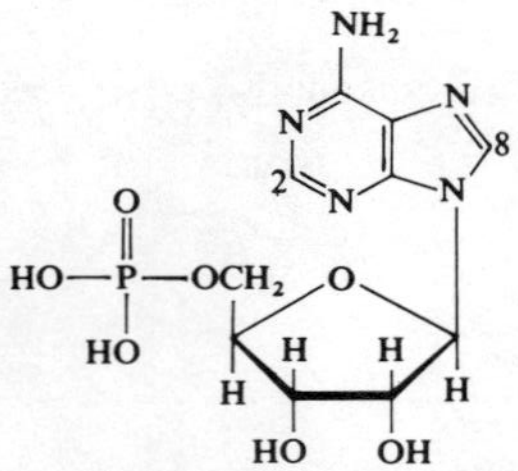 [2,8-^{3}H]Adenosine 5′-monophosphate*	pD 7.8 8.26 8.61	2 8	88 12	D	[23]

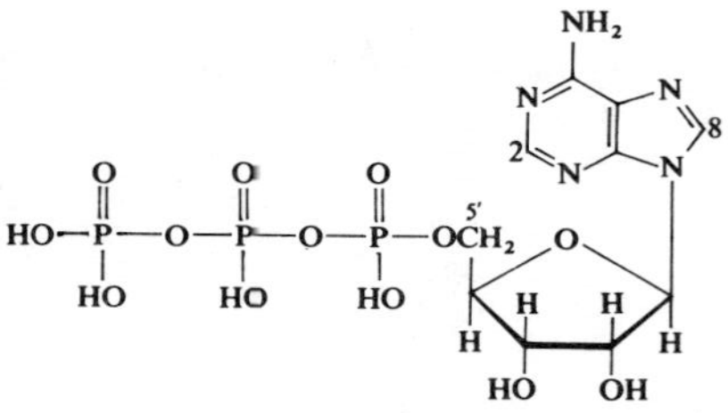 [2,5′,8-^{3}H]Adenosine 5′-triphosphate*	4.23 8.25 8.52	5′ 2 8	62.2 10.6 27.2	B, E	[23]
[2,8-^{3}H]Adenosine 5′-triphosphate*	8.26 8.54	2 8	50 50	D	[23]
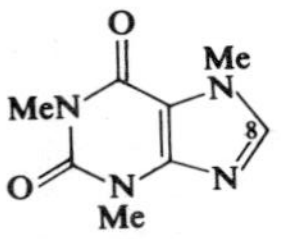 [8-^{3}H]Caffeine	7.89	8	100	E	[49]
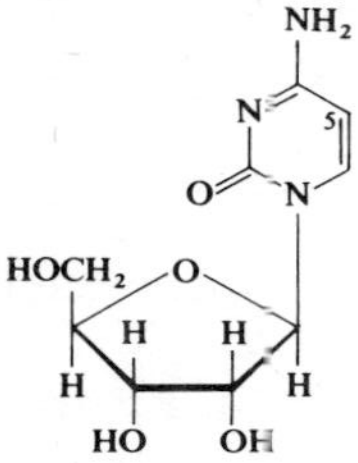 [5-^{3}H]Cytidine	6.09	5	100	D	[23]

Table 26. *(contd.)*

Compound	Chemical shift δ (p.p.m.)	Assignment	Distribution of tritium (%)	Method of preparation	Reference
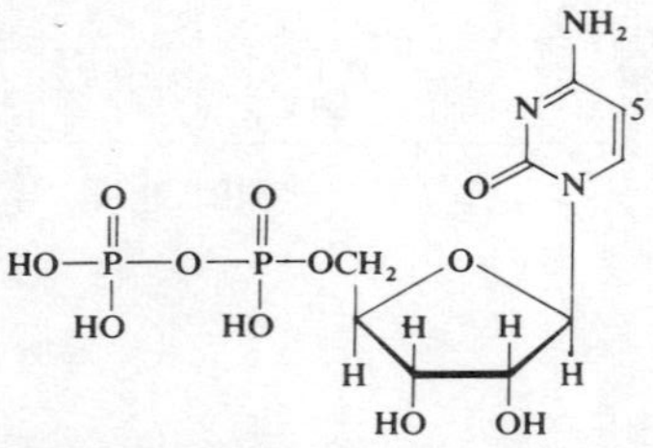 [5-^{3}H]Cytidine 5′-diphosphate*	6.13	5	100	D	[23]
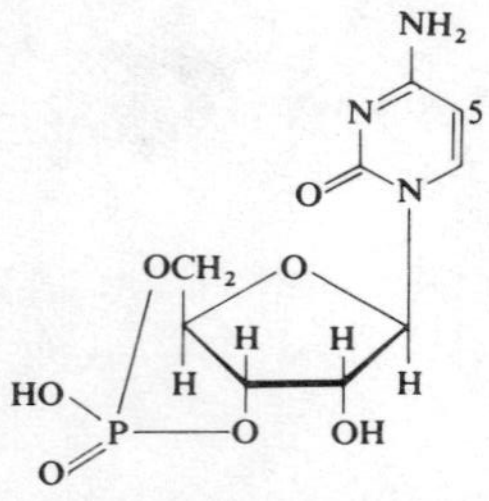 [5-^{3}H]Cytidine cyclic phosphate*	6.05	5	100	D	[23]

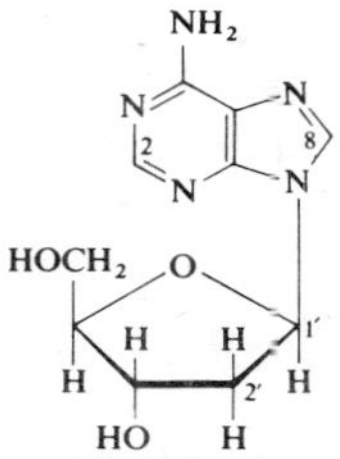

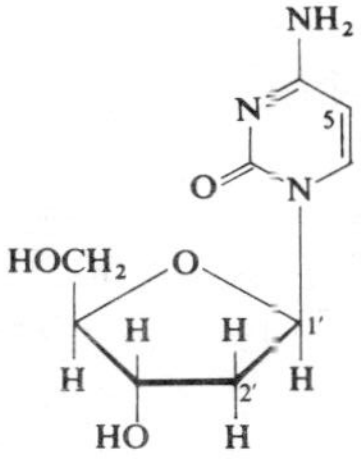

2′-Deoxy[1′,2′,2,8-^{3}H]adenosine	6.5 s+d	1′	21.5	C, D	[23]
	2.58d, 2.59s	2′	29.8		
	8.3	2	21.1		
	8.38	8	27.6		
2′-Deoxy[2,8-^{3}H]adenosine	8.17	2	15	G	[23]
	8.30	8	85		
2′-Deoxy[1′,2′,5-^{3}H]cytidine	6.22 d, 6.23 s	1′	27	C, D	[23]
	2.38 d, 2.39 s	2′	41		
	6.07	5	32		
2′-Deoxy[5-^{3}H]cytidine	6.06	5	100	D	[23]

Table 26. (*contd.*)

Compound	Chemical shift δ (p.p.m.)	Assignment	Distribution of tritium (%)	Method of preparation	Reference
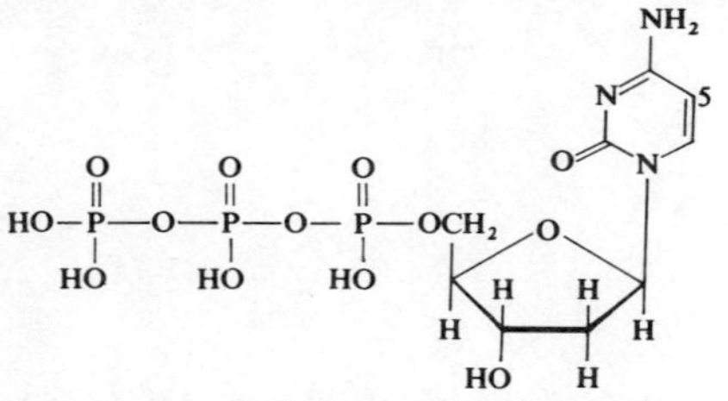 2′-Deoxy[5-^{3}H]cytidine triphosphate*	6.16	5	100	D	[23]
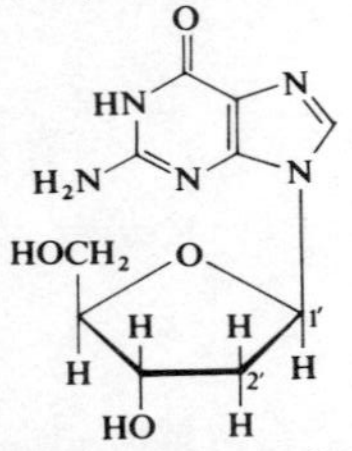 2′-Deoxy[1′,2′-^{3}H]guanosine	6.27 2.48	1′ 2′	33.5 66.5	C	[23]
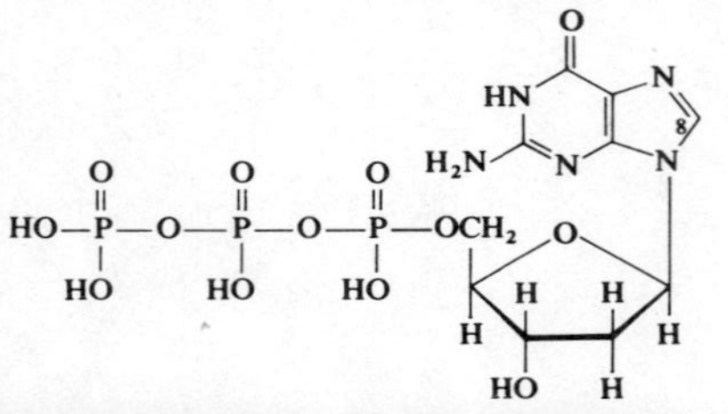 2′-Deoxy[8-^{3}H]guanosine triphosphate*	8.17	8	100	D	[23]

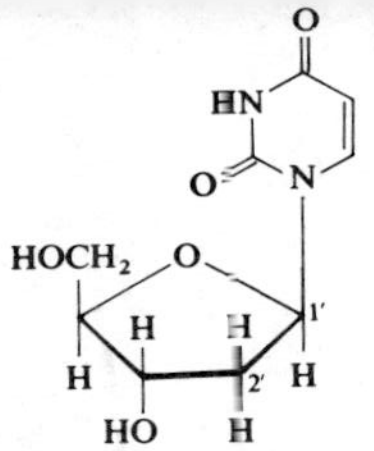 2′-Deoxy[1′,2′-^{3}H]uridine	6.16 2.30	1′ 2′	29 71	C	[23]
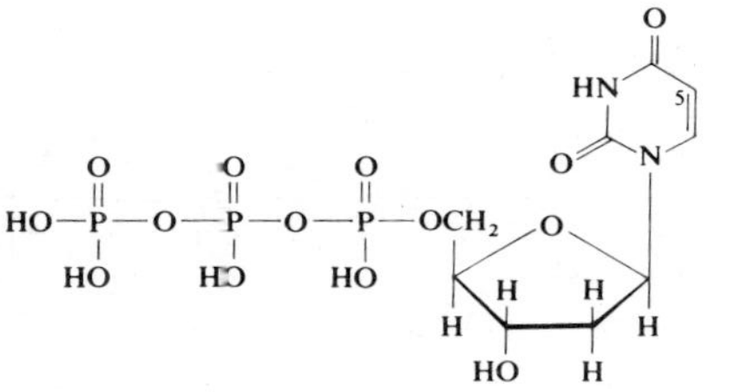 2′-Deoxy[5-^{3}H]uridine triphosphate*	5.97	5	100	D	[23]
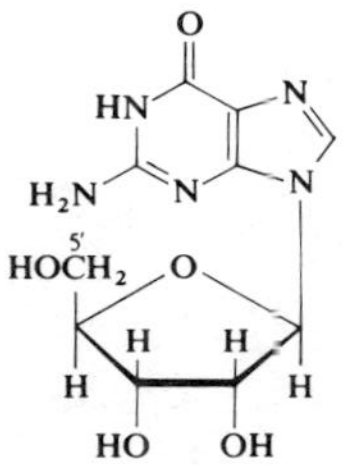 [5′-^{3}H]Guanosine	3.80 3.87	5′	78.2 21.8	B	[23]

Table 26. (*contd.*)

Compound	Chemical shift δ (p.p.m.)	Assignment	Distribution of tritium (%)	Method of preparation	Reference
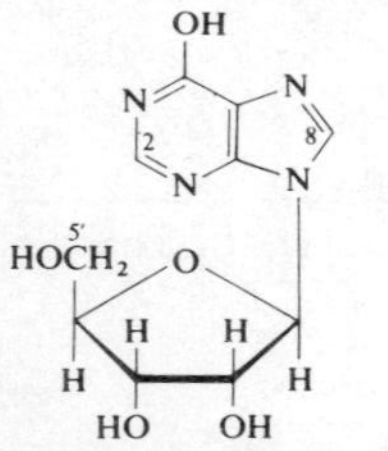 [2,5′,8-^{3}H]Inosine	8.21	2	3.0	B, D	[23]
	3.80, 3.87	5′	33.5, 12.7		
	8.32	8	50.8		
$\beta\gamma$-Methylene[2,8-^{3}H]adenosine triphosphate*	8.26	2	73.8	E	[23]
	8.59	8	26.2		

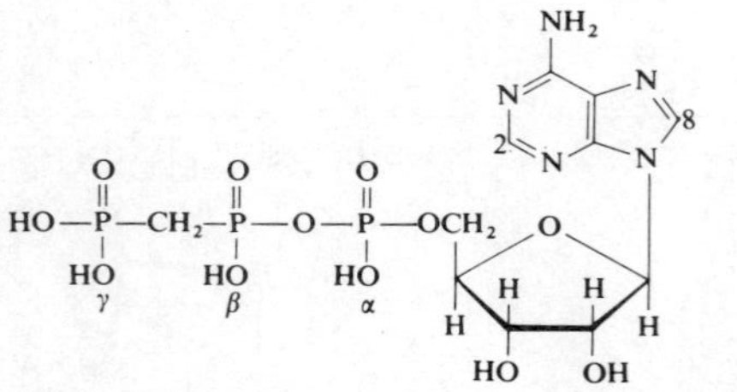

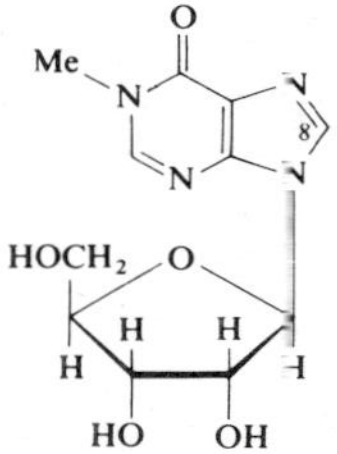 1-Methyl[8-^{3}H]inosine	8.20	8	100	I	[100]
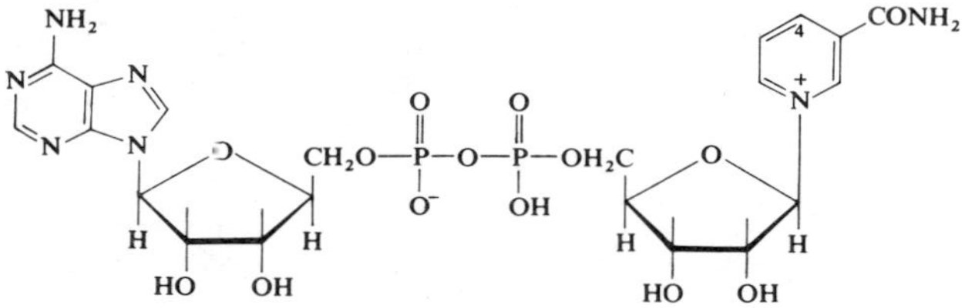 [4-^{3}H]Nicotinamide adenine dinucleotide	8.87	Pyridine-4	100	K	[23]
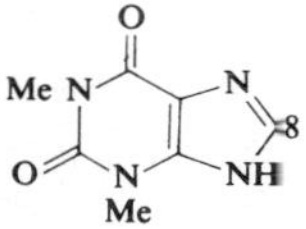 [8-^{3}H]Theophylline	7.85	8	100	E	[49]

Table 26. *(contd.)*

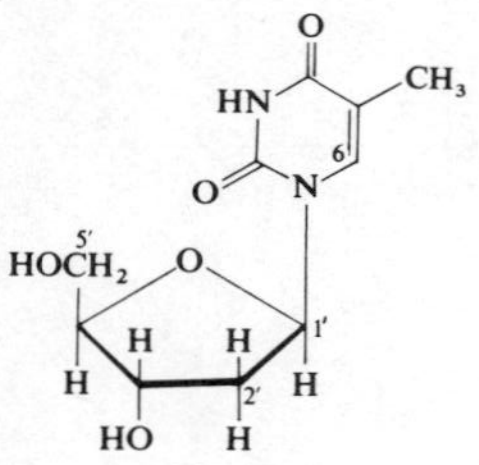

Compound		Chemical shift δ (p.p.m.)	Assignment	Distribution of tritium (%)	Method of preparation	Reference
$[5'\text{-}^3H]$Thymidine		3.72 3.79	5′	100	B	[23]
$[6\text{-}^3H]$Thymidine		7.64	6	100	D	[23]
[*methyl*-^3H]Thymidine	^1H coupled	1.80 s 1.83 d 1.85 t	CT_3 CT_2H CTH_2	100	C	[23]
	^1H decoupled	1.84	Methyl	100		
[*methyl*-^3H]Thymidine	^1H coupled	1.82 s	CT_3	100	A	[23]
[*methyl*, 1′,2′-^3H]Thymidine		6.25 2.33 1.83	1′α 2′α Methyl	13.5 26.0 60.5	C	[23]

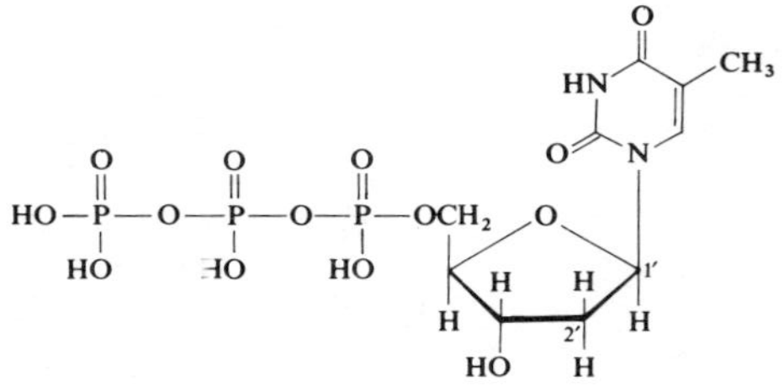

Compound	δ	Assignment	%		Ref.
	^1H coupled spectrum				
	1.88 s	CT_3			
	1.90 d	CT_2H	100		
	1.92 t	CTH_2			
	^1H decoupled spectrum				
[*methyl*-^{3}H]Thymidine triphosphate*	1.89	Methyl	100	C	[23]
[*methyl*, 1′,2′-^{3}H]Thymidine triphosphate* (d_6-DMSO)	6.15 2.02 1.74	1′α 2′α Methyl	14.0 22.7 63.3	C	[23]
[*methyl*, 1′,2′-^{3}H]Thymidine triphosphate*	6.31 2.35 1.86	1′α 2′α Methyl	15.8 22.4 61.8	C	[23]
	^1H coupled spectrum				
[*methyl*-^{3}H]Thymine	1.788 1.814 1.840	CT_3 CHT_2 CH_2T	32 49 19	C	[19]

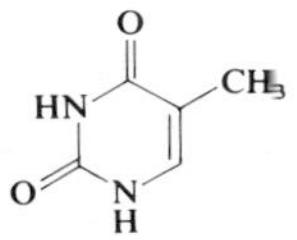

Table 26. (*contd.*)

Compound	Chemical shift δ (p.p.m.)	Assignment	Distribution of tritium (%)	Method of preparation	Reference
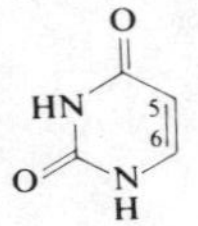 [5-^{3}H]Uracil	5.82	5	93	C	[23]
	7.6	6	7		
[5,6-^{3}H]Uracil	5.82	5	43.9	C	[23]
	7.57	6	56.1		
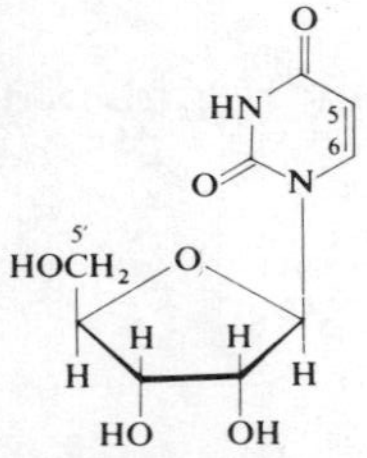 [5-^{3}H]Uridine	5.91	5	100	D	[23]
[6-^{3}H]Uridine	7.86	6	100	D	[23]

Compound	δ (ppm)	Position	%	Method	Ref.
$[5,6\text{-}^3H]$Uridine	5.91 d 7.875 d	5 6	50 50	D	[23]
$[5,5',6\text{-}^3H]$Uridine	5.92 7.88 3.77 3.88	5 6 5′	22.6 32.6 44.8	B, D	[23]
$[5\text{-}^3H]$Uridine triphosphate*	5.98	5	100	D	[23]
$[5,6\text{-}^3H]$Uridine triphosphate*	5.98 7.96 7.98	5 6	47.8 52.2	D	[23]

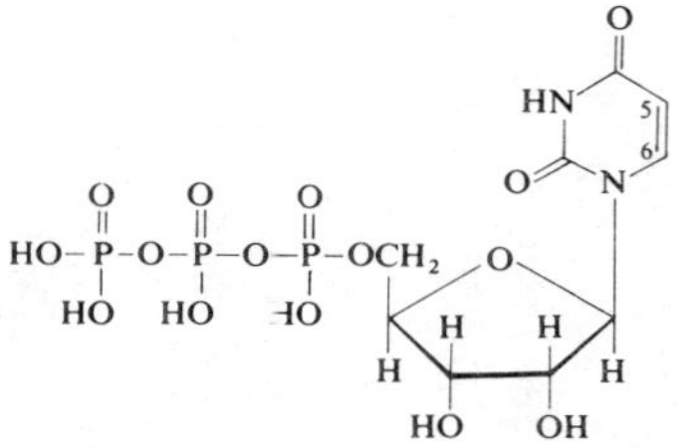

* Ammonium salt.

method of preparation of the labelled thymine was expected to yield a product containing mainly a CHT_2 methyl group together with some CH_2T species. The direct observation of some CT_3 species in addition was initially surprising because aldehyde protons do not exchange. However, the first formed hydrogenolysis product would have a methyl group activated by the heteroaryl nucleus, so that the exchange was most probably a subsequent process. New information concerning the chemistry of a labelling process thus came from the detailed knowledge of the distribution of the tritium in a labelled methyl group. From the signal intensity measurements a value of 61 Ci mmol^{-1} was calculated for the specific activity of the [*methyl*-^{3}H]thymine, closely agreeing with the 58 Ci mmol^{-1} found by direct scintillation counting and u.v. light absorption measurements.

The lines in the ^{1}H-coupled ^{3}H nmr spectrum (Figure 4c) were a trifle broad, the longer range coupling between the methyl group and H-6 not having been resolved. However, further enhancement of the spectrum (at the cost of lower signal-to-noise) revealed the expected doublet splitting of the CT_3 signal ($^4J_{TH}$ = 1.1 Hz) (see Figure 4d) [40].

Tritiated thymidine labelled in the 5′-position of the 2′-deoxyribose moiety shows a spectrum closely similar to that of [6-^{3}H]glucose with signals at $\delta = 3.72$ and 3.79 due to the non-equivalence of the C-5 protons.

Finally it should be mentioned that the chemical shifts of nucleic acids, nucleosides, and the parent bases show (not surprisingly) some dependence on the pH of the solution, as well as on concentration and temperature. As regards adenine, adenosine, and adenylic acid in D_2O, there seems little dependence in the region pD 5.8 to 8.0, but at low pD the chemical shifts of the 2- and 8-hydrogen nuclei change by $+0.2$ p.p.m. or more [23, 98a].

(vi) Tritium labelled steroids

Numerous studies in biochemical and biomedical research require the use of tritium labelled steroids at very high molar specific activity, normally containing at least two tritium atoms per molecule. Such compounds are used in radioimmunoassays and in the study of steroid receptor proteins as well as for studies of metabolism and steroid pathways. For many such uses of tritiated steroids it is not only important to be certain of the positions of the tritium atoms but also of their stereochemistry, i.e. whether the label is alpha or beta. Not surprisingly, a considerable effort has been devoted to these problems and methods have been developed, e.g. stereospecific enzymatic dehydrogenation. However, the methods are all characterized by being demanding in terms of both time and skill. Furthermore, the interpretation of the results depends essentially on the specificity of the chemical and biochemical (enzymatic) procedures adopted and these are not always easy to check. In addition there is usually the problem of tritium loss through unpredictable hydrogen isotope exchange reactions, which need to be taken into account. The summary given in Table 27 of the work done in establishing patterns of labelling in [1,2-^{3}H]testosterone (**XIII**)

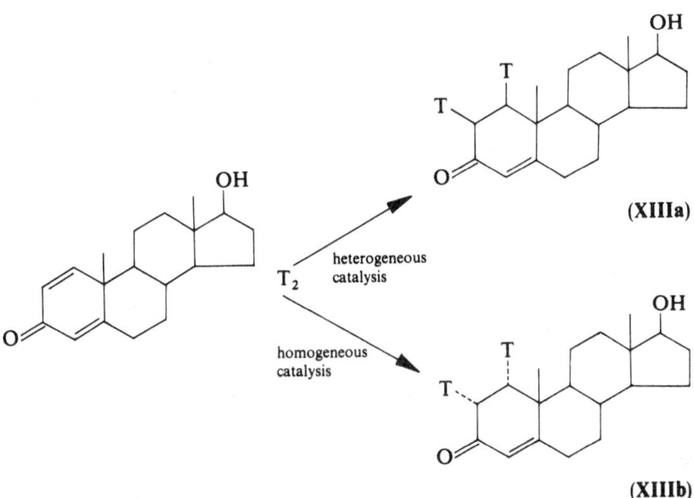

Table 27. Distribution of tritium in [1,2-^{3}H]testosterone prepared by catalytic reduction of 17β-hydroxyandrosta-1,4-dien-3-one. (Reproduced from *Steroids*, **28**, p. 363 (1976) by permission of Holden-Day Inc.)

Catalyst	Position of tritium label	Percentage of tritium	Method of determination	Reference
Pd/C	$1\beta+2\beta$	69	Enzymatic dehydrogenation	[101]
	$2\alpha+2\beta$	43	Alkaline exchange	
	1β	44	Enzyme reaction on preceding product	
	2β, 2α, 1α	25, 18, 13	Calculated	
Pd/C	$1\beta+2\beta$	82	Enzymatic	[102]
	$1\alpha+2\alpha$	18	Calculated	
Pd/C	$2\alpha+2\beta$	50	Alkaline exchange	[21]
	$1\beta+2\beta$	75	Enzymatic dehydrogenation	
	2α	12	Bromination of preceding product	
	2β, 1β, 1α	38, 37, 13	Calculated	
Pd/C	1β, 2β	40, 39	^{3}H nmr spectroscopy	[20]
	1α, 2α	9, 12		
$[Ph_3P]_3RhCl$	$1\beta+2\beta$	13	Enzymatic dehydrogenation	[21]
	$1\alpha+2\alpha$	87	Calculated	
	2α	44	Bromination of above product	
	1α	43	Calculated	
$[Ph_3P]_3RhCl$	1β, 2β	8.5, 7.5	^{3}H nmr spectroscopy	[20]
	1α, 2α	44, 40		

illustrates the considerable effort required and the difficulties in analysing the data for even this relatively simple steroid. Fortunately, however, ^{3}H nmr spectroscopy has simplified the task and provides an invaluable tool for establishing both the positions and stereospecificity of the tritium labelling [20, 32].

[1,2-^{3}H]Testosterone is normally prepared by catalytic reduction of 17β-hydroxyandrosta-1,4-dien-3-one with tritium gas [1]. Depending upon the choice of catalyst, either the [1β,2β-^{3}H]form (**XIIIa**) or the [1α,2α-^{3}H]form (**XIIIb**) is produced preferentially, though not exclusively. The distribution of labelling was established by previous investigators in the following manner. Alkaline exchange studies gave the proportion of the label located in the 2-position, and enzymatic dehydrogenation gave the proportion present in both the 1β- and 2β-positions together. Bromination of the enzymatic dehydrogenation product gave the proportion of tritium in the 2α-position. Finally the value for the 1α-position could be obtained by difference (Table 27) [21].

In sharp contrast to these lengthy procedures, the ^{3}H nmr spectrum provided the required distribution data for each position directly. Thus for [1β,2β-^{3}H]testosterone prepared using a palladium-on-charcoal catalyst, the ^{3}H nmr (proton decoupled) spectrum (Figure 27a) shows two strong doublet signals, from

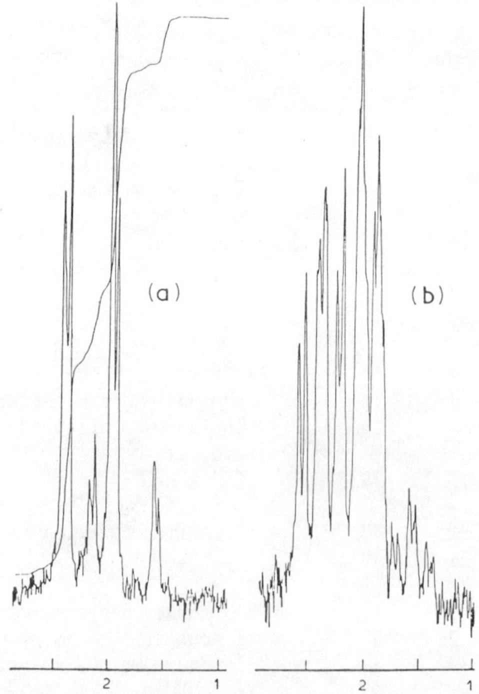

Figure 27. ^{3}H Nmr spectra of [1β,2β-$^3H_2$]testosterone (**XIIIa**) in d_6-DMSO (a) with ^{1}H-decoupling and (b) with ^{1}H-coupling. (Reproduced from *Steroids*, 28, p. 364 (1976) by permission of Holden-Day, Inc.)

the 1β- and 2β-tritons, with $J_{TT} = 5.5$ Hz (vicinal, equatorial–axial coupling), indicating predominantly [$1\beta,2\beta$-$^3H_2$]labelling. In addition, there were two weak doublet signals from the $1\alpha,2\alpha$-tritons, showing that some hydrogenation had also occurred on the α-face of the substrate molecule. The proton-coupled ^{3}H nmr spectrum (Figure 27b) proved the assignments [20]. The product (**XIIIb**), formed when the homogeneous catalyst tris-(triphenylphosphine)rhodium(I) chloride was used, gave a ^{3}H nmr spectrum (Figure 28) which was complementary to that just described, with strong doublets from the 1α- and 2α-tritons and weak ones from tritons in the 1β- and 2β-positions, again showing the catalyst to be incompletely stereospecific. The corresponding undecoupled ^{3}H nmr spectrum (Figure 28b) confirms, by the high field triplet of doublets, that the 1α-triton is indeed axial, being coupled equally to the geminal 1β-proton and the axial 2β-proton (large triplet splitting), as well as to the equatorial 2α-triton (small doublet splitting of the triplet lines). The data are given fully in Table 28. Tritiated progesterone, prepared in the same manner, gives a similar distribution of tritium (Table 28), and neither of the two catalysts leads to a fully stereospecifically labelled product.

One example of a non-specific labelling reaction following the catalytic halogen–tritium replacement in 7-bromopregnenolone has already been discussed

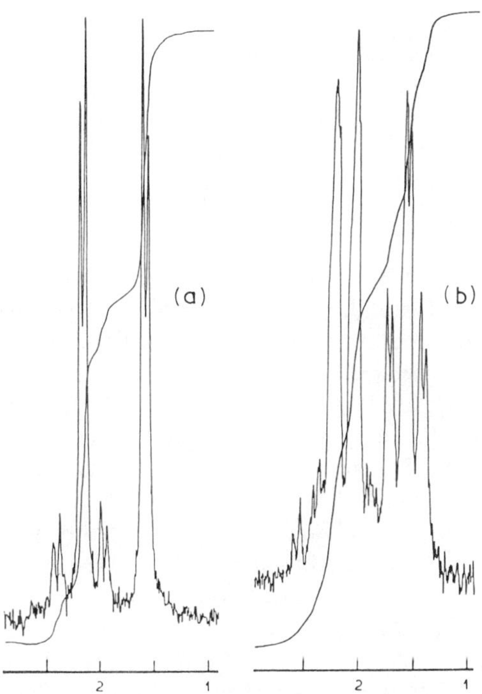

Figure 28. ^{3}H Nmr spectra of [$1\alpha,2\alpha$-$^3H_2$]testosterone (**XIIIb**) in d_6-DMSO (a) with ^{1}H-decoupling and (b) with ^{1}H-coupling. (Reproduced from *Steroids*, **28**, p. 365 (1976) by permission of Holden-Day, Inc.)

Table 28. Chemical shifts δ (p.p.m), integrated intensities (%), assignments and coupling information (Hz) from ^{3}H nmr spectra of tritiated testosterone and progesterone at high dilution in d_6-DMSO at 25°C. (Reproduced from *Steroids*, **28**, p. 367 (1976) by permission of Holden-Day Inc.)

Compound	Shift	Intensity (**XIIIa**)	Intensity (**XIIIb**)	Assign-ment	Apparent splitting*	$\lvert J(\mathrm{T,T})\rvert$
Testosterone	1.58	9	44	1α	t of d 15	
						4.8
	2.14	12	40	2α	d 18	
	1.96	40	8.5	1β	d of dd 14, 3	
						5.5
	2.38	39	7.5	2β	t of d 16	
Progesterone	1.54		37.5	1α	t of d 14.5	
						4.5
	2.12		37.5	2α	d 18	
	1.89		12	1β	d	
						4.9
	2.35		13	2β	d	

* d = doublet, t = triplet, dd = double doublet.

(see page 28). Another interesting example occurs in the preparation of tritiated estriol. Bromination of the phenolic ring of estriol followed by catalytic halogen–tritium replacement led to only 80 per cent. of the tritium being located in the ring at the expected 2- and 4-positions (53 and 27 per cent., respectively), and the remainder was located at the 6- and 9-benzylic positions (2 and 18 per cent., respectively). These two latter positions are those labelled when estriol is subjected to exchange in solution with tritium gas and a palladium-on-charcoal catalyst [1, 20].

Several steroids have been tritiated by a combination of labelling methods. An example is [2,4,6,7-^{3}H]estradiol prepared by reduction of the $\Delta^{6,7}$-unsaturated precursor followed by bromination of the 2,4-positions and then catalysed halogen–tritium replacement. The ^{1}H-decoupled ^{3}H nmr spectrum of this resultant [2,4,6,7-^{3}H]estradiol (Figure 29) shows clearly that the reactions have been completely specific; 36 per cent. of the tritium is located in the aromatic ring, at H-2 ($\delta = 6.46$) and H-4 ($\delta = 6.32$), with the remaining 64 per cent. being located at H-6 ($\delta = 2.54$) and H-7 ($\delta = 1.03$) in the ratio 2:3. An even more impressive example is that of [2,4,6,7,16,17-^{3}H]estradiol, prepared by the same principle using a $\Delta^{6,7,16,17}$-unsaturated precursor in the first tritiation stage. The ^{3}H nmr spectrum of the product (Figure 30) again shows directly that the labelling is exactly as expected.

Another but slightly different use of ^{3}H nmr spectroscopy in the analysis of steroids was made when preparing [^{3}H]betamethasone. This compound is very difficult to separate and distinguish from [^{3}H]dihydrobetamethasone—both compounds exhibit the same ultraviolet light absorption characteristics and virtually the same chromatography patterns. Yet the ^{3}H nmr spectra (Figure 31)

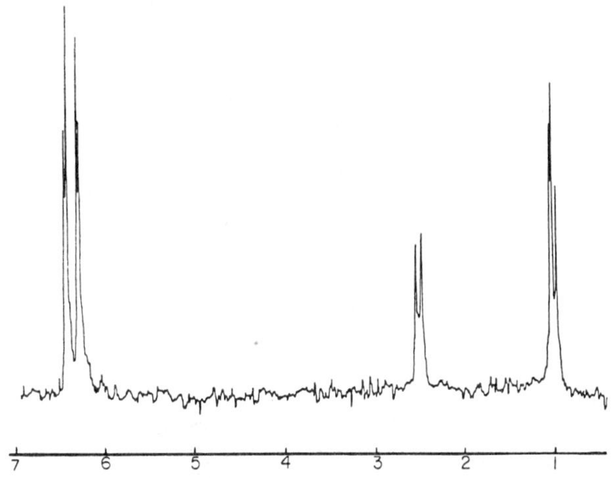

Figure 29. ^{3}H Nmr spectrum (^{1}H-decoupled) of [2,4,6,7-^{3}H]estradiol in d_6-DMSO. (Reproduced from *Steroids*, **29**, p. 563 (1979) by permission of Holden-Day, Inc.)

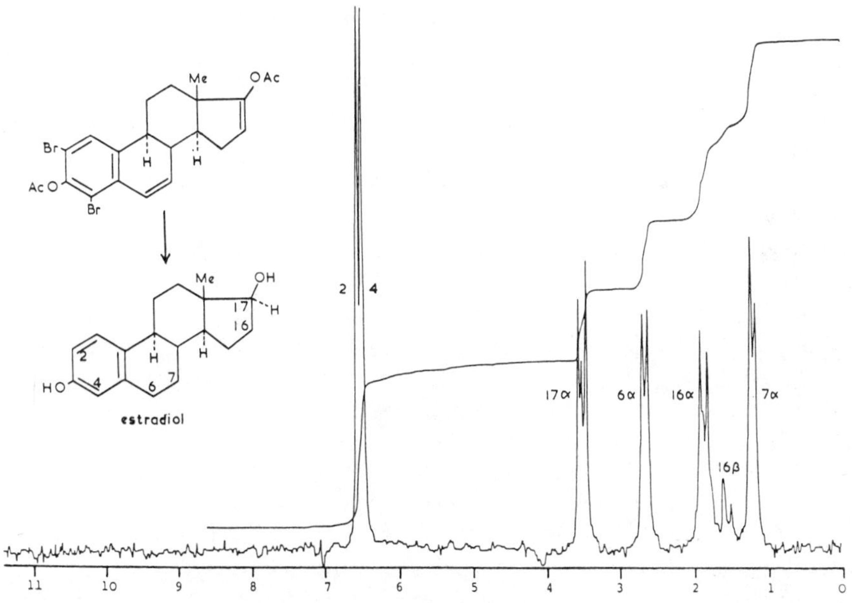

Figure 30. ^{3}H Nmr spectrum (^{1}H-decoupled) of [2,4,6,7,16,17-^{3}H]estradiol in d_6-DMSO

are quite distinct and serve to distinguish between the two compounds and to provide the proportions of mixtures of the two.

The chemical shifts and distribution patterns in a variety of tritiated steroids are given in Table 29.

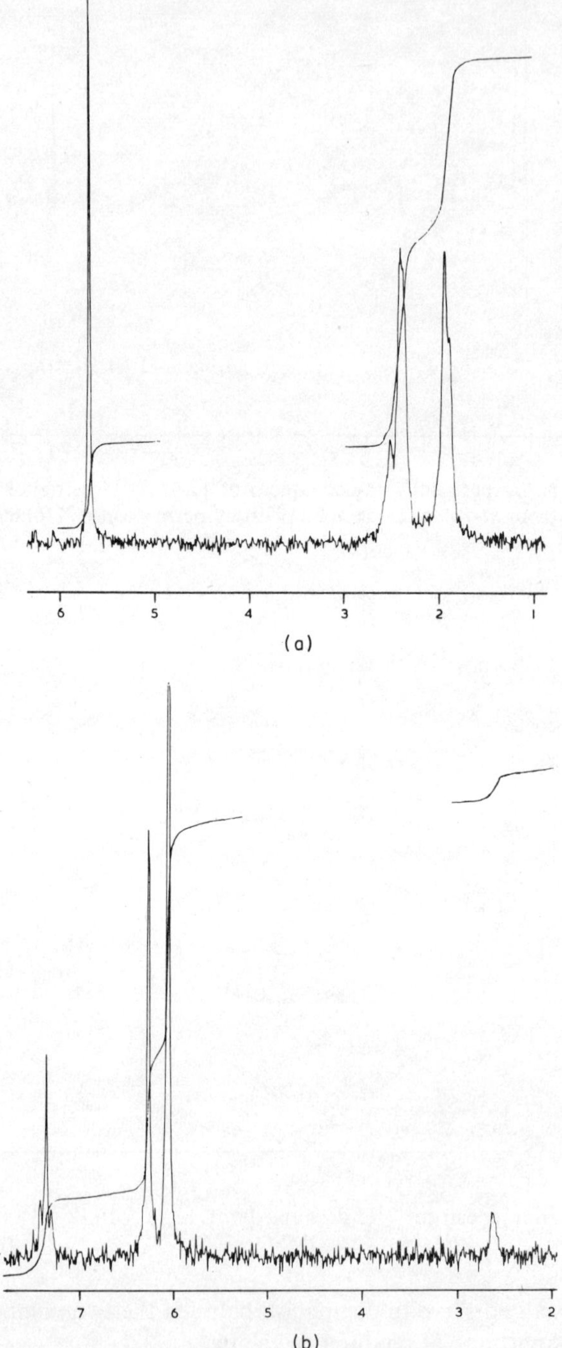

Figure 31. ^{3}H Nmr spectra (^{1}H-decoupled) of (a) [^{3}H]dihydrobetamethasone and (b) [^{3}H]betamethasone, in d_6-DMSO.

Table 29. Chemical shifts and distribution of labelling for some tritium labelled steroids in d_6-DMSO except where otherwise indicated.

Compound	Chemical shift δ (p.p.m.)	Assignment	Distribution of tritium (%)	Method of preparation	Reference
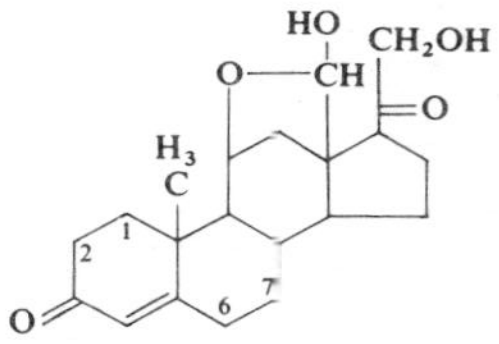					
	1.58	1α	47.5		
	2.14	2α	48.6	C	[23]
[1,2-^{3}H]Aldosterone	2.41	2β	3.8		
	1.57	1α	20.6		
	1.85	1β (or 7β)	5.2		
[1,2,6,7-^{3}H]Aldosterone	2.15	2α and 6α	41.6	C	[23]
	2.38	2β	7.8		
	1.05	7α	24.7		
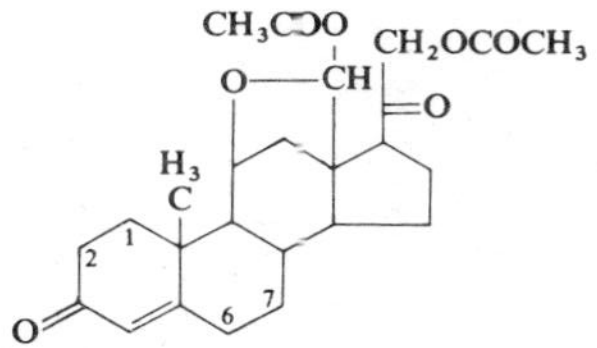					
	1.58	1α	24.0		
	1.82	1β (or 7β)	4.0		
	2.1	2α and 6α	40.0	C	[23]
	2.37	2β	7.5		
[1,2,6,7-^{3}H]Aldosterone 18,21-diacetate	1.04	7α	24.5		

Table 29. (*contd.*)

Compound	Chemical shift δ (p.p.m.)	Assignment	Distribution of tritium (%)	Method of preparation	Reference
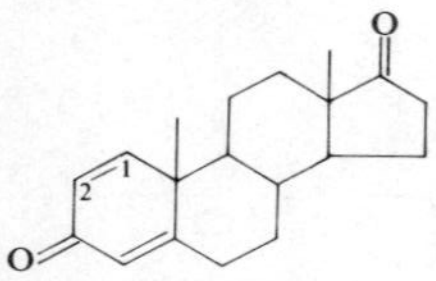					
	7.23	1	17		
[1,2-^{3}H]Androsta-1,4-diene-3,17-dione	6.15 ($^3J_{TH}$ = 12.6 Hz)	2	83	C	[103, 104]
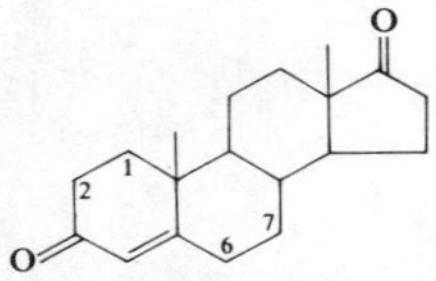	1.56	1α	25.1		
	2.12	2α	20.5		
	2.24	6α	22.8	C	[23]
	0.98	7α	25.1		
[1,2,6,7-^{3}H]Androst-4-ene-3,17-dione	1.88	7β	6.4		
	0.98	7α	38	D	[20]
[7(*n*)-^{3}H]Androst-4-ene-3,17-dione	1.87	7β	62		
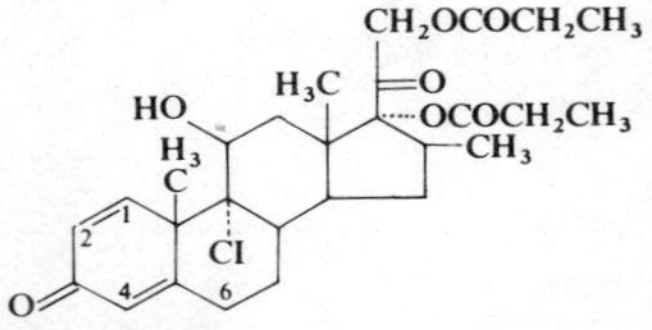	7.34	1	24.8		
	6.27	2	54.7	C	[23]
[1,2-^{3}H]Beclomethasone 17α,21-dipropionate	2.61	6	20.5		

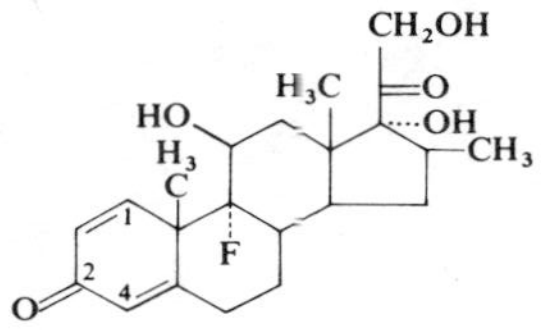	7.33	1	10.0		
	6.25	2	37.3	C	[23]
	6.04	4	42.7		
[1,2,4-^{3}H]Betamethasone	2.61	6	10.0		
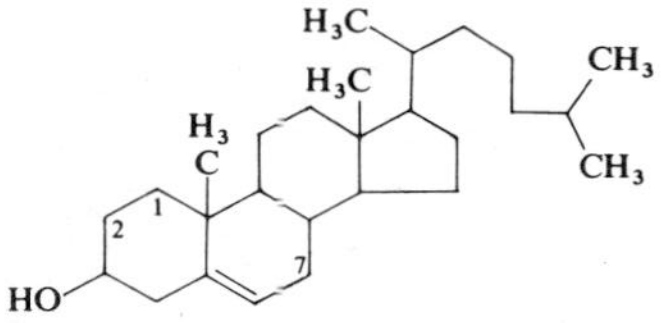	1.02	1α (coupled to 1β)	46.4		
	1.03	1α (coupled to 2α)			
	1.47	1β (coupled to 2α)	16.8	C	[23]
	1.48	1β (coupled to 1α)			
[1α,2α(*n*)-^{3}H]Cholesterol (in $CDCl_3$)	1.81	2α (coupled to 1α and 1β)	36.8		
	1.48	7α	46		
[7(*n*)-^{3}H]Cholesterol (in $CDCl_3$)	1.93	7β	54	D	[23]

Table 29. *(contd.)*

Compound	Chemical shift δ (p.p.m.)	Assignment	Distribution of tritium (%)	Method of preparation	Reference
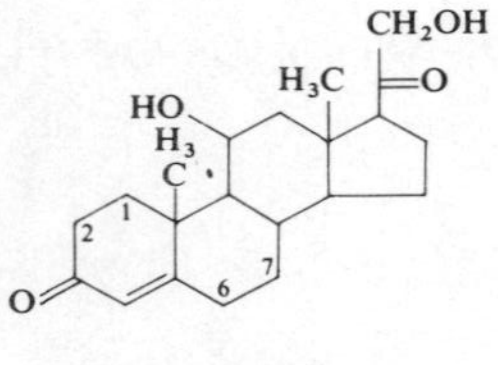 [1,2,6,7-^{3}H]Corticosterone	1.74	1α	21.6	C	[23]
	2.16	$2\alpha + 6\alpha$	40.5		
	1.91	$1\beta + 7\beta$	4.0		
	2.38	$2\beta + 6\beta$	7.3		
	0.95	7α	26.6		
[1,2,6,7-^{3}H]Cortisol	1.74	1α	13.3	C	[23]
	1.87	1β or 7β	7.9		
	2.15	2α or 6α	39.4		
	2.36	2β	16.7		
	0.94	7α	22.6		

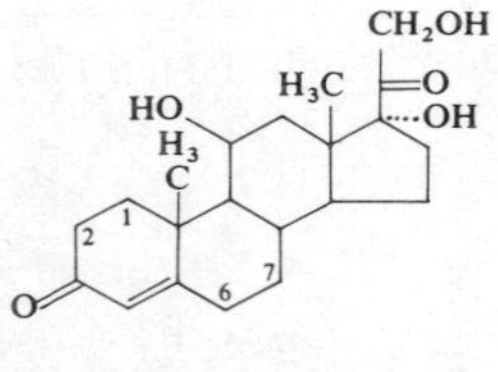

[*cyclopropyl*-^{3}H]Cycloartenol (in d_6-benzene)	0.168 0.438 0.456	exo (endo D) endo (exo D) endo (exo H)	65.4 14.1 20.5	B	[105]
Dehydro[1,2 6,7-^{3}H]epiandrosterone	0.94 1.61 1.73 1.65 5.31 1.55 2.01	1α 1β 2α 2β 6α 7α 7β	26 5 9 10 15.5 19 15.5	C	[20]
Dehydro[7-^{3}H]epiandrosterone	1.57 2.02	7α 7β	39 61	D	[20]
Dehydro[7(*n*)-^{3}H]epiandrosterone 3-acetate 17-ethyleneketal	1.61 2.05 2.29 1.86	7α 7β 4α 4β	31.9 50.6 9.6 7.8	D	[23]

Table 29. (*contd.*)

Compound	Chemical shift δ (p.p.m.)	Assignment	Distribution of tritium (%)	Method of preparation	Reference
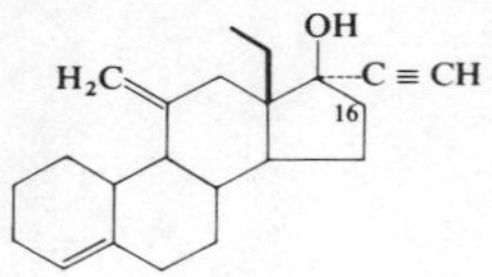	2.09	16α (16β)	18.0		
	2.12	16α	32.8		
	1.90	16β (16α)	18.0	J	[106]
	1.94	16β	31.2		
(in $CDCl_3$)	2.29	16α (16β)	17.6		
	2.33	16α	36.1		
	2.04	16β (16α)	17.6	J	[106]
[16-^{3}H]Desogestrel	2.07	16β	28.6		
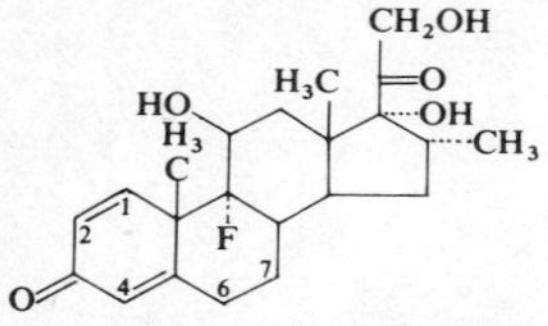	7.34	1	91.5	C	[23]
[1,2-^{3}H]Dexamethasone	6.19	2	8.5		
	7.33	1	8		
	6.25	2	42	C	[23]
[1,2,4-^{3}H]Dexamethasone	6.04	4	50		
[1,2,4,6,7-^{3}H]Dexamethasone	7.33	1	12.0		
	6.25	2	18.7		
	6.04	4	18.3	C	[23]
	2.60	6	26.1		
	1.76	7	24.9		

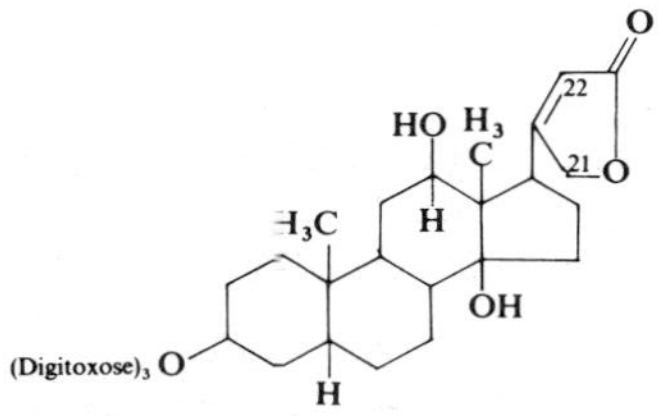

[21,22-^{3}H]Digoxin	4.82 4.89 5.85	21 21 22	33.1 29.6 37.3	J	[23]
Dihydro[1,2-^{3}H]beclomethasone 17α,21-dipropionate	2.11 2.49	1 2	48 52	C	[23]

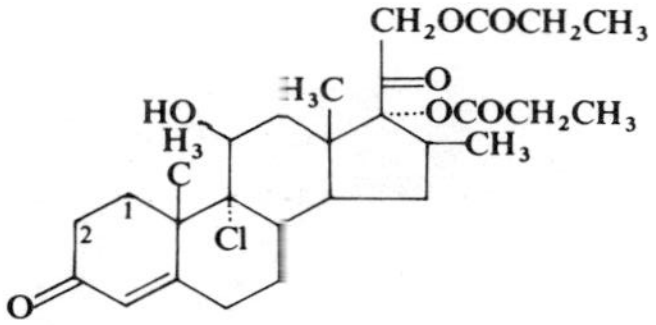

Table 29. *(contd.)*

Compound	Chemical shift δ (p.p.m.)	Assignment	Distribution of tritium (%)	Method of preparation	Reference
Dihydro[1,2-^{3}H]betamethasone 17α,21-dipropionate	1.95	1	37.4	C	[23]
	2.42	2	44.3		
	5.71	4	18.3		
5α-Dihydro-19-nor[1,2-^{3}H]testosterone (in d_6-benzene)	0.81	1α	16.25	C	[106]
	0.80	1α (2α)	28.75		
	1.77	1β	6.25		
	1.76	1β (2β)	8.75		
	2.29	2α	1.25		
	2.28	2α (1α)	28.75		
	1.87	2β	1.25		
	1.85	2β (1β)	8.75		
5α-Dihydro-19-nor[1,2-^{3}H]testosterone (in $CDCl_3$)	1.21	1α	52.4	C	[106]
	1.18	1α (2α)	5.8		
	2.24	1β	19.4		
	2.35	2α (1α)	5.8		
	1.57	$^3H^1HO$	16.5		

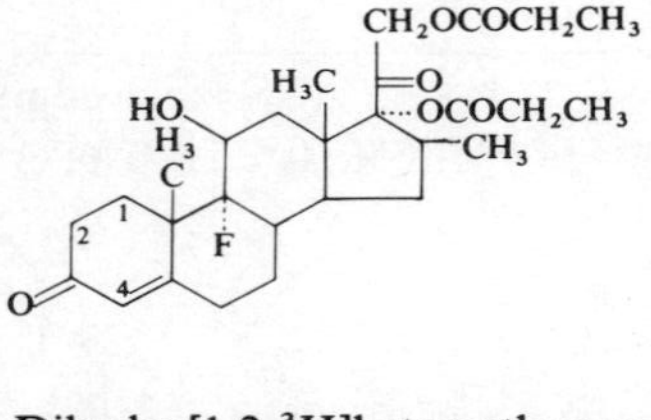

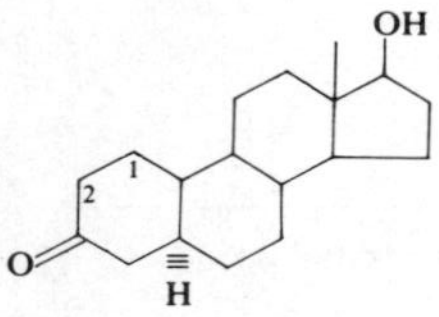

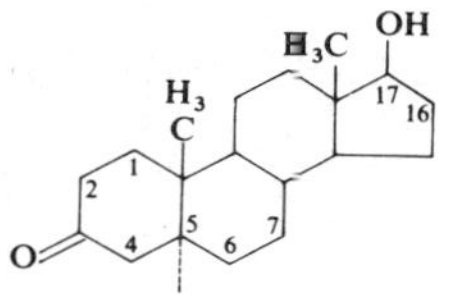

5α-Dihydro[1α,2α(n)-^{3}H]testosterone	1.22	1α coupled to 2β	20.5	C	[23]
	1.23	1α coupled to 2α	21.9		
	1.24	1α	15.1		
	2.06	2α coupled to 1α	21.9		
	2.40	2β coupled to 1α	20.5		
5α-Dihydro[4,5-^{3}H]testosterone	1.40	5α	58.0	C	[104]
	1.86	4α	26.9		
	2.29	4β	15.1		
5α-Dihydro[1,2,4,5,6,7-^{3}H]testosterone	1.23	1α, 6α	27.6	C	[23]
	2.05	2α	15.7		
	2.40	2β	1.9		
	1.86	4α	17.8		
	2.31	4β	1.9		
	1.40	5α	18.9		
	0.80	7α	16.2		
5α-Dihydro[1,4,5,6,7,16,17-^{3}H]testosterone	1.24	1α, 6α	32.0	C	[23]
	1.86	4α }	13.2		
	1.81	16α }			
	1.40	5α	21.3		
	0.80	7α	17.3		
	3.40	17α	16.2		

Table 29. (*contd.*)

Compound	Chemical shift δ (p.p.m.)	Assignment	Distribution of tritium (%)	Method of preparation	Reference
	6.56	2	32		
	6.48	4	15	C,D	[20]
	2.65	6α	24		
[2,4,6,7-^{3}H]Estradiol	1.18	7α	28		
	6.46	2 }	36		
	6.32	4 }			
[2,4,6,7-^{3}H]Estradiol (in d_6-benzene)	2.54	6α	25.4	C,D	[32]
	1.03	7α	38.6		
	6.53	2	16.3		
[2,4,6,7,16,17-^{3}H]Estradiol	6.48	4	14.1		
	2.65	6α	14.5		
	1.18	7α	18.1	C,D	[23]
	1.85	16α	17.0		
	1.53	16β	3.6		
	3.48	17α	16.3		
	2.64	6α	50	C	[20]
[6,7-^{3}H]Estradiol	1.20	7α	50		

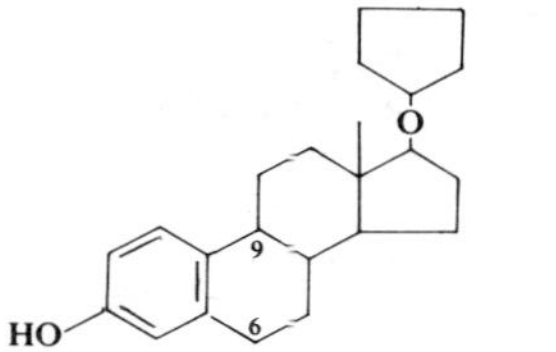

Compound	δ (ppm)	Position	%	Method	Ref.
[6,9-^{3}H]Estradiol 17-cyclopentyl ether (in $CDCl_3$)	2.80	6β	33	E	[106]
	1.25	7α	1		
	2.13	9α	66		
[2,4(*n*)-^{3}H]Estriol	6.57	2	53	D	[20]
	6.51	4	27		
	2.68	6α	2		
	2.06	9	18		
[2,4,6,9-^{3}H]Estriol	6.56	2	26	D,E	[20]
	6.51	4	19		
	2.65	6α	31		
	2.10	9	24		
[6,9-^{3}H]Estriol	2.65	6α	61	E	[20]
	2.10	9	39		
[2,4,6,7-^{3}H]Estrone	6.61	2 }	49	D,C	[20]
	6.54	4 }			
	2.72	6α	24		
	1.30	7α	27		

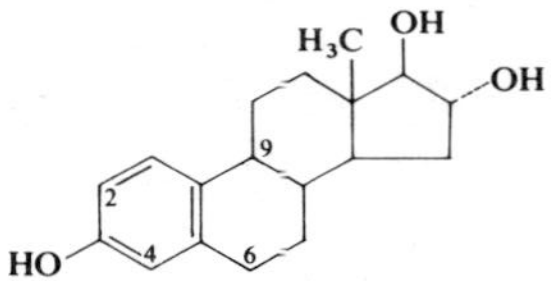

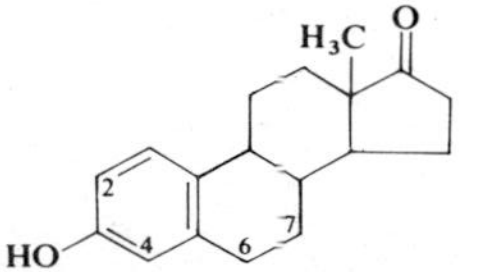

Table 29. (*contd.*)

Compound	Chemical shift δ (p.p.m.)	Assignment	Distribution of tritium (%)	Method of preparation	Reference
β-D-Glucuronide–O (H_3C, O, 9, 6) [6,9-^{3}H]Estrone-β-D-glucuronide	2.79 2.16	6α 9	59.2 40.8	E	[23]
$NaOSO_2O$ (H_3C, O, 9, 6) [6,9-^{3}H]Estrone sulphate	2.77 2.15	6α 9	50 50	E	[20]
(CH_2OH, H_3C, O, CH_2CH_3, O, 6, 7) 16α-Ethyl-21-hydroxy-19-nor[6,7-^{3}H]pregn-4-ene-3,20-dione	2.40 2.23 0.97 1.72	6α 6β 7α 7β	14.5 34.8 15.9 34.8	C	[23]

Compound	δ (ppm)	Position	%		Ref.
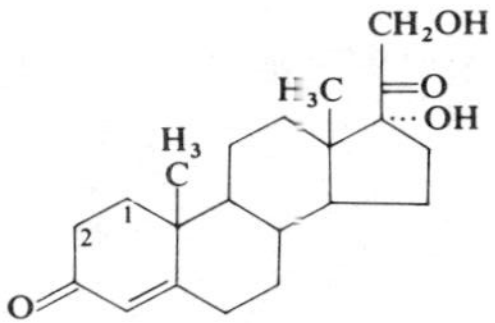 17α-Hydroxy-11-deoxy[1,2-^{3}H]corticosterone	1.56 1.90 2.12 2.35	1α 1β 2α 2β	42.1 9.0 40.7 8.1	C	[104]
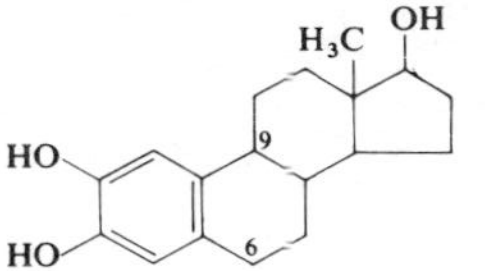 2-Hydroxy[6,9-^{3}H]estradiol (in d_5-pyridine)	2.76 2.80 2.14	6α } 6β } 9	58 42	E	[23]
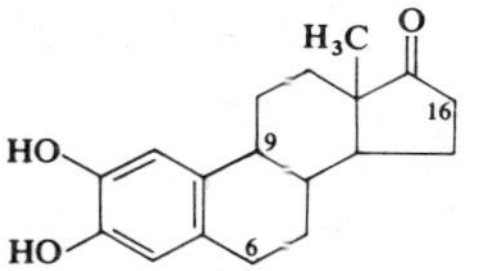 2-Hydroxy[6,9-^{3}H]estrone	2.60 2.64 2.06 1.32	6 } 6 } 9 16	52.6 42.7 4.7	E	[23]

Table 29. (*contd.*)

Compound	Chemical shift δ (p.p.m.)	Assignment	Distribution of tritium (%)	Method of preparation	Reference
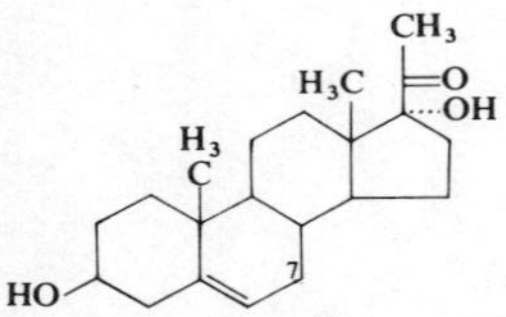 17α-Hydroxy[7-^{3}H]pregnenolone	1.52	7α	43	D	[23]
	1.92	7β	57		
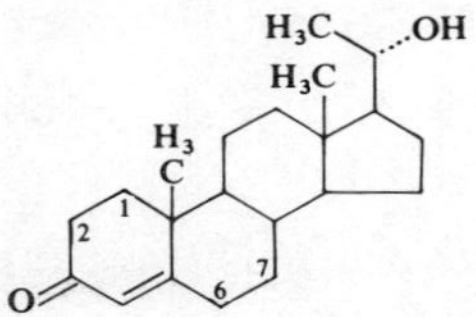 20α-Hydroxy[1,2,6,7-^{3}H]pregn-4-en-3-one	1.57	1α	24.8	C	[23]
	1.94	1β	3.3		
	2.13	2α	16.3		
	2.34	2β	2.0		
	2.21	6α	13.7		
	2.38	6β	2.3		
	0.94	7α	34.0		
	1.76	7β	3.6		
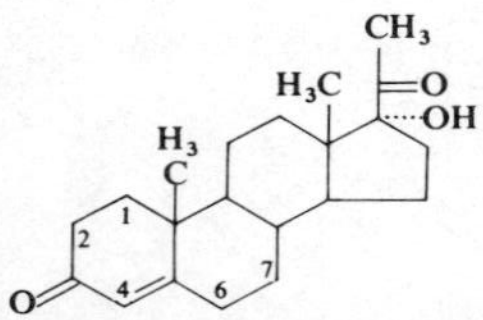 17α-Hydroxy[1,2,6,7-^{3}H]progesterone	1.57	1α	15.5	C	[23]
	1.92	1β	11.7		
	2.12	2α	14.9		
	2.35	2β, 6β	17.2		
	2.21	6α	12.3		
	0.95	7α	17.5		
	1.75	7β	11.0		

7α-Methyl[16-^{3}H]norethinodrel (in $CDCl_3$)	2.30 2.00	16α 16β	50 50	I	[106]
D-[16-^{3}H]Norgestrel (in $CDCl_3$)	2.30 2.08	16α 16β	53 47	J	[106]
19-Nor[7-^{3}H]testosterone	0.93 1.73	7α 7β	86 14	C	[23]

Table 29. (*contd.*)

Compound	Chemical shift δ (p.p.m.)	Assignment	Distribution of tritium (%)	Method of preparation	Reference
$O-\overset{O}{\overset{\Vert}{C}}-(CH_2)_8CH_3$	2.44	6α	1.5		
	2.43	6α (7α)	10.3		
	2.23	6β	2.9		
	2.21	6β (7β)	22.8		
(in $CDCl_3$)	1.03	7α	10.3	C	[106]
	1.02	7α (6α)	10.3		
	1.80	7β	19.1		
	1.78	7β (6β)	22.8		
	2.40	6α	1.2		
19-Nor[6,7-^{3}H]testosterone decanoate	2.39	6α (7α)	11.0		
	2.23	6β	6.1		
	2.22	6β (7β)	27.6		
	0.95	7α	5.5	C	[106]
	0.84	7α (6α)	11.0		
	1.73	7β	9.8		
	1.72	7β (6β)	27.6		
OH, $C\equiv CH$, H_2C, 16	2.35	16α	38.8		
	2.32	16α (16β)	17.4		
	2.11	16β	26.4	I	[106]
3-Oxo[16-^{3}H]desogestrel (in $CDCl_3$)	2.08	16β (16α)	17.4		

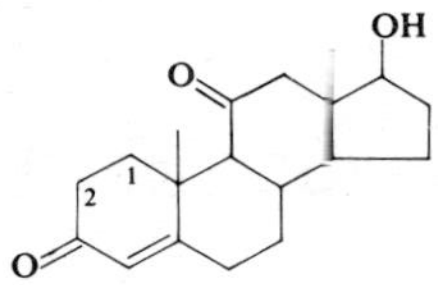 11-Oxo[1,2-^{3}H]testosterone	1.59 2.40 2.09 2.54	1α 1β 2α 2β	26.8 23.2 22.3 27.7	C	[23]
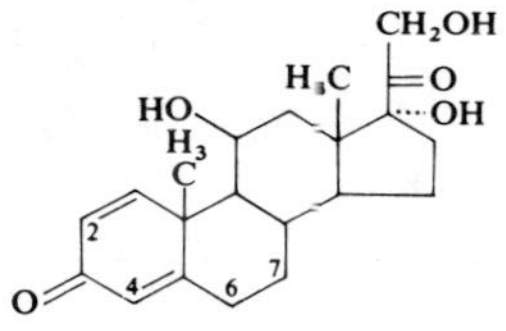 [2,4,6,7-^{3}H]Prednisolone	6.21 5.97 2.28 2.50 0.99 2.03	2 4 6α 6β 7α 7β	10.6 17.9 21.2 7.9 26.0 16.3	C	[23]
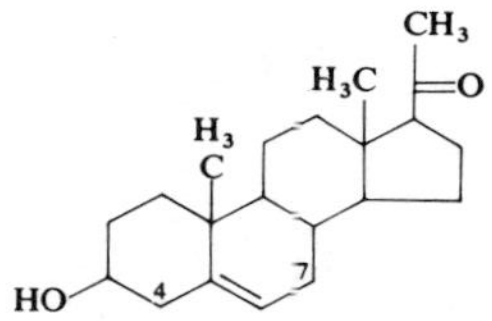 Δ^5-[7(*n*)-^{3}H]Pregnenolone	1.49 1.90 2.13	7α 7β 4α	52 28 20	D	[20]

Table 29. *(contd.)*

Compound	Chemical shift δ (p.p.m.)	Assignment	Distribution of tritium (%)	Method of preparation	Reference
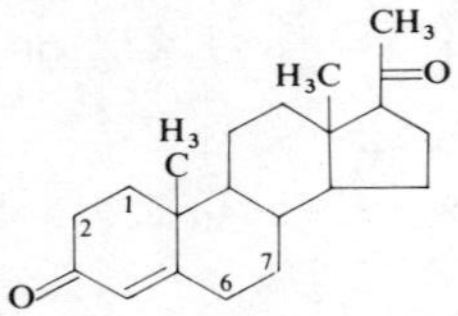 [1,2-^{3}H]Progesterone	1.54	1α	37.5	C	[20]
	1.89	1β	12		
	2.12	2α	37.5		
	2.35	2β	13		
[1,2,6,7-^{3}H]Progesterone	1.57	1α	29	C	[23]
	2.13	2α	21.2		
	2.21	6α	21.2		
	0.94	7α	28.6		
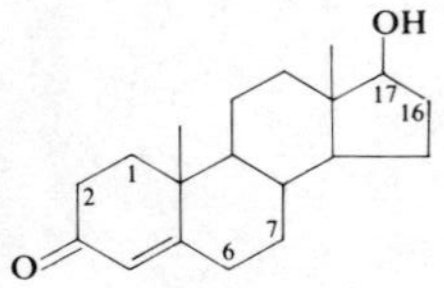 [1α,2α(*n*)-^{3}H]Testosterone	1.58	1α	44	C	[20]
	2.14	2α	40		
	1.96	1β	8.5		
	2.38	2β	7.5		
[1α,2α(*n*)-^{3}H]Testosterone (in d_6-benzene)	1.13	1α }	86	C	[32]
	2.17	2α }			
	1.37	1β }	14		
	2.07	2β }			

[1β,2β(*n*)-^{3}H]Testosterone	1.58	1α	9	C	[20]
	2.14	2α	12		
	1.96	1β	40		
	2.38	2β	39		
[1β,2β(*n*)-^{3}H]Testosterone (in d_6-benzene)	1.16	1α	6.15	C	[32]
	2.173	2α	6.15		
	1.388	1β	61.0		
	2.078 } 2.088 }	2β	26.7		
[1,2,6,7-^{3}H]Testosterone	1.55	1α	19.6	C	[23]
	1.7–1.9	1β, 7β	10.3		
	2.12	2α	19		
	2.3–2.4	2β, 6β	6		
	2.20	6α	20.1		
	0.87	7α	25		
[1,2,6,7,16,17-^{3}H]Testosterone	1.56	1α	17.2	C	[23]
	1.94	1β	4.1		
	2.14	2α	3.6		
	2.40	2β	1.5		
	2.20	6α	15.2		
	2.32	6β	3.6		
	0.87	7α	15.7		
	1.73	7β	3.0		
	1.81	16α	17.8		
	3.40	17α	18.3		
[7-^{3}H]Testosterone	0.88	7α	45.7	D	[23]
	1.74	7β	54.3		

Table 29. *(contd.)*

Compound	Chemical shift δ (p.p.m.)	Assignment	Distribution of tritium (%)	Method of preparation	Reference
[1,2,4,5-^{3}H]Tetrahydrodexamethasone	1.73	1α	23.7	C	[23]
	2.05	2α	23.7		
	1.86	4α	27.7		
	1.44	5α	24.9		

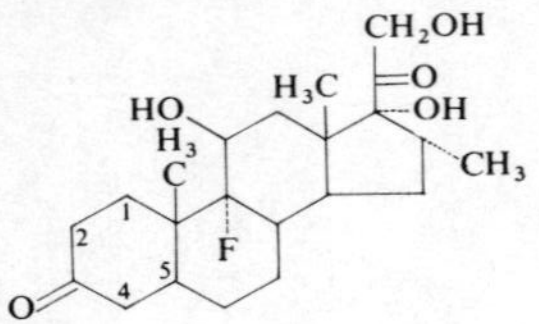

(vii) Tritium labelled aromatic and heterocyclic compounds

This group of tritium labelled compounds which has been studied by tritium nuclear magnetic resonance spectroscopy, with the results listed in Table 30, comprises many different kinds of molecules. These vary from relatively simple benzenoid compounds such as [^{3}H]phenol to much more complex molecules such as [^{3}H]actinomycin D: actinomycin D is a known potent inhibitor of DNA-dependent RNA synthesis. The whole group includes numerous drugs [107] and neurochemicals [49]; compounds which exert their effects in biological systems by interacting with specific receptor sites on the target proteins. Tritium labelled compounds at very high molar specific activities (at least 10 Ci mmol^{-1}) have played and continue to play a vital role in the understanding of drug–receptor interactions and are used frequently as receptor-specific ligands. As with the many other studies with tritium labelled compounds [1], a precise knowledge of the distribution pattern of the label (Table 30) is often essential in order to interpret results which involve any loss of the label from the ligand. In numerous examples, the preparative method leads to completely specific labelling. For example, the thermal decarboxylation of ibotenic acid (**XIV**) in tritiated water yields specifically labelled muscimol. Again for catalysed reactions, by suitable choice of catalyst and optimization of the experimental conditions, a remarkably high degree of specificity of labelling can be achieved.

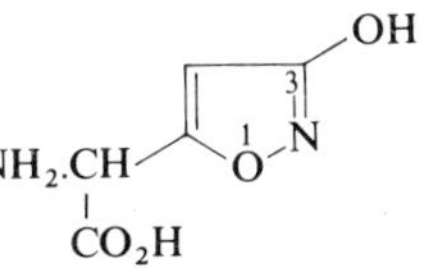

OH
3
1 N
NH_2.CH O
CO_2H

(**XIV**)

A number of important observations and conclusions can be made from the ^{3}H nmr spectral data for a selection of the tritiated compounds in Table 30.

Amphetamine sulphate, subjected to homogeneous catalysed exchange with heptafluorobutyric acid and tritiated water [107], shows limited labelling in the aliphatic side chain. In addition there is steric hindrance to labelling in the ortho-positions of the phenyl ring. Similarly, steric effects inhibit the labelling at the 1- and 10-positions of phenanthridine, tritiated by platinum catalysed exchange in tritiated water at 125°C (for up to 65 hours) [108]. The ^{3}H nmr spectrum of [^{3}H]dihydroalprenolol, an important β-adrenoceptor blocking agent, prepared by palladium catalysed reduction of alprenolol with tritium gas, shows that there has been uneven addition of tritium across the ethylenic bond. The cause is now known to be vinylic hydrogen–tritium exchange, discussed further in Chapter 3. The tritiation of ethyl β-carboline-3-carboxylate (**XV**), an important ligand for benzodiazepine receptor studies, obtained by catalysed halogen–tritium replace-

Table 30. Chemical shifts and distribution of labelling for some tritium labelled aromatic and heterocyclic compounds

Compound	Solvent	Chemical shift δ (p.p.m.)	Assignment	Distribution of tritium (%)	Method of preparation	Reference
[*methyl*-^{3}H]Acetophenone	$CDCl_3$	2.54	Methyl	100	I	[23]
N-Acetyl-5-hydroxy[2,4,6-^{3}H]tryptamine	d_6-DMSO	7.07 6.86 6.63 2.67	2 4 6 α-CH_2	42 31 19 8	D	[40,54]
[^{3}H]Actinomycin	CD_3OD	2.14 2.55	4-methyl 6-methyl	56.2 43.8	E	[54]

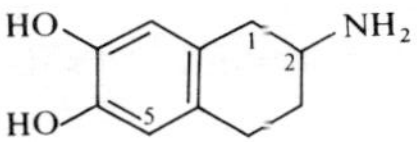

2-Amino-6,7-dihydroxy-1,2,3,4-tetrahydro[^{3}H]naphthalene	D_2O	2.71	1ax.	26	E	[49]
		3.01	1eq.	17		
		1.79	3ax.	6		
		2.15	3eq.	4		
		2.73	4	44		
		6.71	5,8	3		
[G-^{3}H]Amphetamine sulphate	D_2O	7.40	3,5	56	H	[107]
		7.33	4	33		
		2.86	CH_2	8		
		1.24	CH_3	3		
[G-^{3}H]Aniline	Neat	6.35	2,6	79	J	[94]
		7.05	3,5	<1		
		6.69	4	21.1		
[G-^{3}H]Aniscle	Neat	6.77	2,6	64.4	J	[94]
		6.82	4	35.6		
[^{3}H]Baclofen	D_2O	2.73s	2	31	C	[40]
		2.72d	2 coupled to 3			
		3.40s	3	69		
		3.40d	3 coupled to 2			

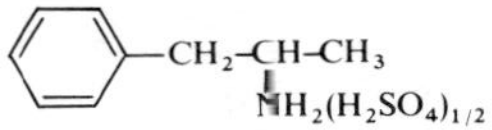

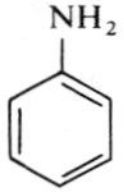

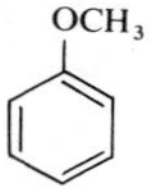

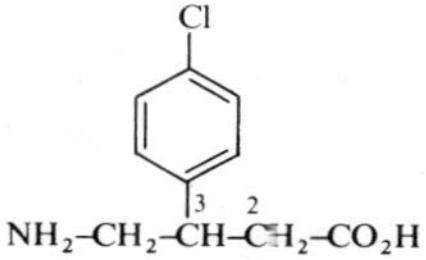

Table 30. (*contd.*)

Compound	Solvent	Chemical shift δ (p.p.m.)	Assignment	Distribution of tritium (%)	Method of preparation	Reference
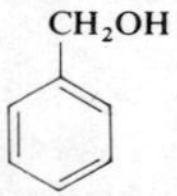 CH_2OH (structure) [^{3}H]Benzyl alcohol	$CDCl_3$	4.0	CH_2	100	J	[93]
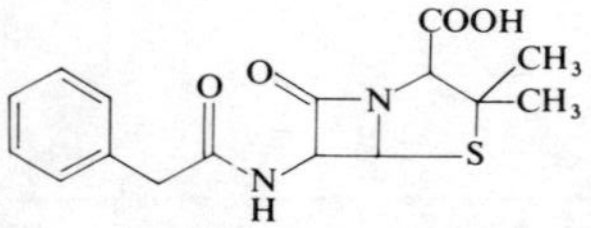 (structure: COOH, CH_3, CH_3, O, O, N, S, NH) [^{3}H]Benzylpenicillin *N*-ethylpiperidine salt	CD_3OD	7.2–7.5 3.57	Phenyl Benzylic	96 4	D	[23]
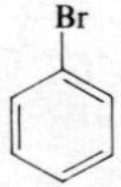 Br (structure) Bromo[^{3}H]benzene	Neat	7.29 6.94 7.01	2,6 3,5 4	26.4 < 1 73.6	G, J	[94,95]

Compound	Solvent	δ	Position	%		Method	Ref.
Chloro[^{3}H]benzene	Neat	7.14 7.00 6.97	2,6 3,5 4	— — —		G, J	[95]
	5% in Cyclohexane	7.25 7.16 7.11	2,6 3,5 4	— — —		G,J	[95]
	Neat	7.05 7.02 7.18	2,6 4 3,5	(a) 39 20 41	(b) 57 (39+20) 43	J	[93]
1-Chloro[^{3}H]naphthalene	CCl_4	7.43 7.26 7.59 7.76 8.10	2,7 3,6 4 5 8	(a) 34 32 23 11 0	(b) 30 8 10 25 26	G, J	[93]
$CH_3NHC(=NCN)NHCH_2CH_2SCH_2$-(5-methylimidazol-4-yl) [*N-methyl*-^{3}H]Cimetidine	d_6-DMSO	2.68	*N*-methyl	100		A	[49]
$C_6H_5CH{=}CHCO_2H$ [α-^{3}H]Cinnamic acid	$CDCl_3$	6.55	α-CH	100		H	[109]

Table 30. (*contd.*)

Compound	Solvent	Chemical shift δ (p.p.m.)	Assignment	Distribution of tritium (%)	Method of preparation	Reference
[*phenyl*-4-^{3}H]Clonidine	D_2O	7.45	Phenyl-4	100	D	[49]
2,4-Diamino[7-^{3}H]pteridine-6-carboxylic acid	D_2O	9.05	7	100	E	[53]
[*N-methyl*-^{3}H]Diazepam	d_6-DMSO	3.26(s) 3.28(d)	CT_3 CT_2H	94 6	A	[19,49]

Compound	Solvent	δ (ppm)	Position	%		Method	Ref.
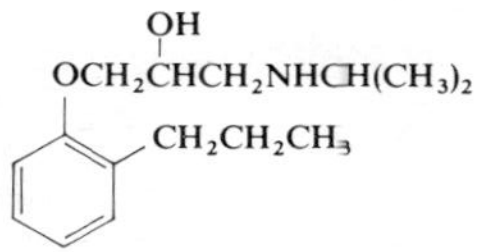 OCH$_2$CH(OH)CH$_2$NHCH(CH$_3$)$_2$ / CH$_2$CH$_2$CH$_3$ (structure)	d_6-DMSO	2.54 m	Propyl-1	7		C	[49]
		1.53 m	Propyl-2	29			
		0.86 m	Propyl-3	64			
[^{3}H]Dihydroalprenolol	d_6-DMSO	2.52 m	Propyl-1	5		C,D	[49]
		1.50 m	Propyl-2	22			
		0.83 m	Propyl-3	50			
		6.93	Ring 4,6	23			
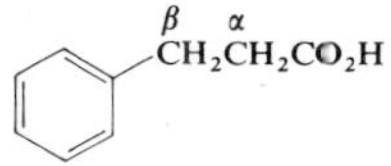 $\overset{\beta}{C}H_2\overset{\alpha}{C}H_2CO_2H$ (structure)				(a)	(b)		
[^{3}H]Dihydrocinnamic acid	CDCl$_3$	2.74 d	α-CH$_2$	50	19.5	C	[96]
		2.96 d	β-CH$_2$	50	47		
		7.7	Aromatic	—	33.5		
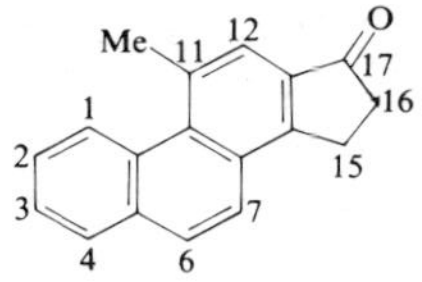 (structure: Me, 1, 2, 3, 4, 6, 7, 11, 12, 15, 16, 17, O)				(a)	(b)		
[G-^{3}H]15,16-Dihydro-11-methyl cyclopenta[a]phenanthren-17-one	CDCl$_3$	2.84	16-CH$_2$	19.0	24.1	G	[110]
		3.17	11-CH$_3$	23.1	19.4		
		3.46	15-CH$_2$	14.8	12.5		
		7.72	2,3	17.4	17.9		
		7.86	12	1.9	2.7		
		7.93	6	8.7	6.2		
		7.99	7	3.2	3.4		
		8.03	4	11.9	13.8		
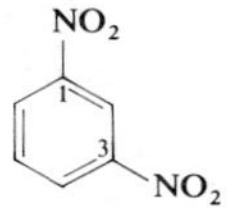 1,3-Dinitro[^{3}H]benzene (structure: NO$_2$, 1, 3, NO$_2$)	d_6-acetone	8.9	2	93		I	[111]
		8.7	4(6)	7			

Table 30. (*contd.*)

Compound	Solvent	Chemical shift δ (p.p.m.)	Assignment	Distribution of tritium (%)	Method of preparation	Reference
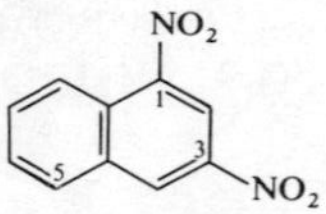 1,3-Dinitro[^{3}H]naphthalene (structure: NO_2 at 1, NO_2 at 3; position 5 marked)	d_6-DMSO	9.35 8.9	2 4	15 85	I	[112]
	Dioxan (with external D_2O as lock)	8.19 7.80 7.94	5 6 7	10 45 45	J	[112]
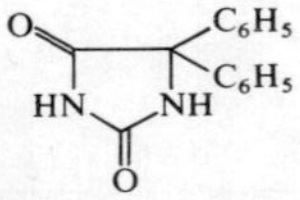 [^{3}H]Diphenylhydantoin (structure: C_6H_5, C_6H_5, HN, NH, O)	d_6-DMSO	7.43 7.38	3,5 4	61 39	H	[107]
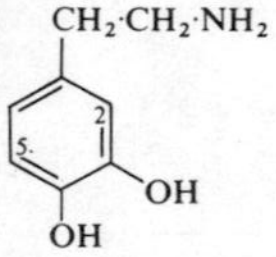 [^{3}H]Dopamine hydrochloride (structure: $CH_2{\cdot}CH_2{\cdot}NH_2$, OH, OH; positions 2, 5 marked)	D_2O	2.82 3.18	Ar-CH_2 *N*-CH_2	48 52	C	[49]
	D_2O	6.87 6.93 6.79	2 5 6	44 21 35	H	[49]
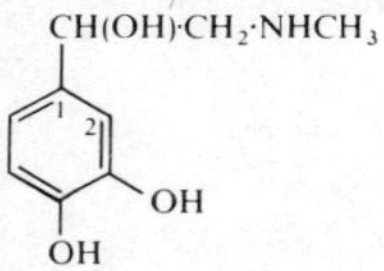 DL-[^{3}H]Epinephrine (structure: $CH(OH){\cdot}CH_2{\cdot}NHCH_3$, OH, OH; positions 1, 2 marked)	D_2O/ CD_3CO_2D	3.27 4.90	*N*-CH_2 *O*-CH	61 39	C	[49]
	D_2O	4.89	*O*-CH	100	B	[49]

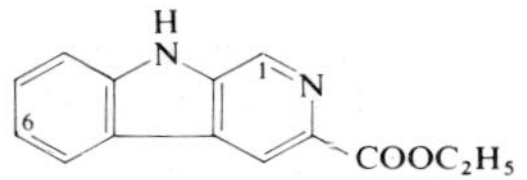

Compound	Solvent	δ (ppm)	Position	%	Method	Ref.
Ethyl β-[^{3}H]carboline-3-carboxylate	d_6-DMSO	9.01 7.39	1 6	(a) (b) 6.2 43 93.8 57	D	[54]
[^{3}H]Fenvalerate	C_6D_6	6.91	Phenyl-4	100	D	[23]
[*N*-methyl-^{3}H]Flunitrazepam	d_6-DMSO	3.36 s 3.39d	CT_3 CT_2H	94 6	A	[19,49]
Fluoro[^{3}H]benzene	Neat	6.86 7.06 6.89	2,6 3,5 4	— — —	G,J	[95]

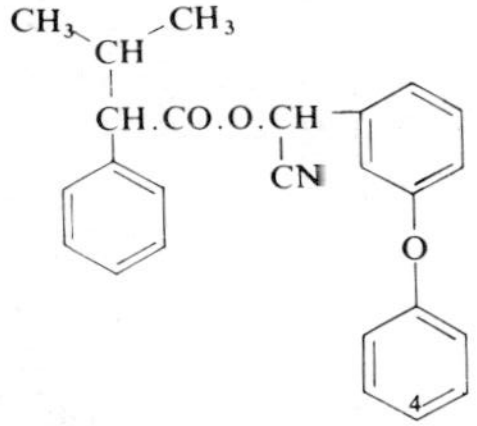

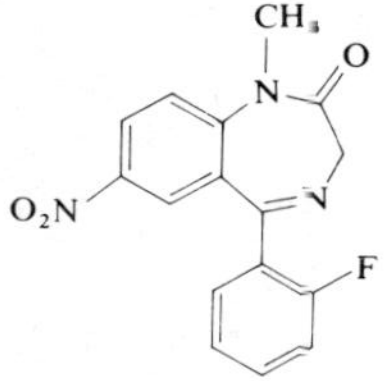

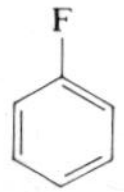

Table 30. (*contd.*)

Compound	Solvent	Chemical shift δ (p.p.m.)	Assignment	Distribution of tritium (%)	Method of preparation	Reference
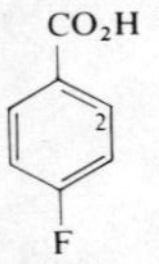 4-Fluoro[2-^{3}H]benzoic acid	d_6-DMSO	8.04	2	100	D	[104]
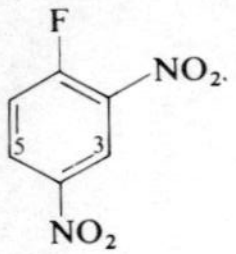 1-Fluoro-2,4-dinitro[3,5-^{3}H]benzene	CCl_4	9.32 9.00	3 5	50 50	D	[4]
4-Fluoro-3-nitro[^{3}H]-phenylazide (FNPA)	$CDCl_3$	7.306 7.307	6 coupled to 2 and F 6 coupled to F	53	D	[23]
		7.753 7.754	2 coupled to 6 and F 2 coupled to F	47		

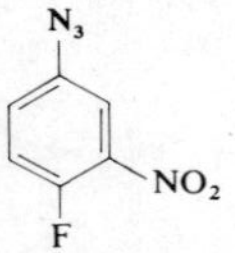

Compound	Solvent	δ	Position	%		Method	Ref.
				(a)	(b)		
	D$_2$O	6.82	3′, 5′	73.1	42.5	D	[53]
		8.63	7	2.8	32.0		
		4.53	9-methylene	24.1	25.5		
[^{3}H]Folic acid, potassium salt	D$_2$O	8.63	7	59		E	[53]
		4.53	9-methylene	41			
[2,5-^{3}H]Histamine dihydrochloride	d_6-DMSO	9.00	2	29		D	[49]
		7.53	5	71			
5-Hydroxy[G-^{3}H]tryptamine creatinine sulphate	d_6-DMSO	7.31	2	19		G	[49]
		7.12	4	21			
		6.89	6	22			
		7.43	7	22			
		3.07	CH_2-N	16			
[^{3}H]Imipramine	d_6-DMSO	7.17	3,9	80		D	[49]
		3.06	10,11	20			
	CDCl$_3$	7.23	3,9	22.6		C,D	[23]
		3.14	10,11	77.4			
	CF_3CO_2D	7.28	1,2,3,7,8,9	55		H	[107]
		7.48	4,6	45			

Table 30. *(contd.)*

Compound	Solvent	Chemical shift δ (p.p.m.)	Assignment	Distribution of tritium (%)	Method of preparation	Reference
$HNCH_2CH_2\overset{+}{C}(=NH_2)OCH_2CH_2SO_3^-$ (2,4-dinitrophenyl; NO_2, NO_2) Isoethionyl 3-(*N*-2,4-dinitro[3,5-^{3}H]phenyl)aminopropioimidate	d_6-DMSO	8.83 8.29	3 5	43.3 56.7	D	[23]
(CO_2H; 6, 2, NH) [^{3}H]Isoguvacine hydrobromide	D_2O	3.38 3.87	2-CH_2 6-CH_2	56 44	B	[49,54]
(N) [G-^{3}H]Isoquinoline	Neat	9.52 8.84 7.54 7.62 7.50 7.41 7.78	1 3 4 5 6 7 8	14 14 12 10 18 17 15	G	[108]

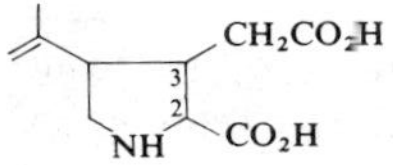

Compound	Solvent	δ	Position	%		
$[^3H]$Kainic acid	D_2O	(a) (b) 1.7 1.84 3.36 ca 3.86 4.35 4.75 4.90 5.0 5.10	 Methyl $CH_2-CO_2^-$ 5-CH_2 $CH-CO_2^-$ $=CH_2$	(i) (ii) 75 32 — 45 — 18 — trace 25 5	H	[49]
2,6-[G-^{3}H]Lutidine	10% in CCl_4	2.39 6.82 7.33	Methyls 3,5 4	76 7 17	G	[108]
	D_2O	6.82 8.52	3′,5′ 7	(i) (ii) 37 46.4 63 53.6	D	[53]
	D_2O	8.59	7	100	E	[53]
$[^3H]$Methotrexate, sodium salt	d_6-DMSO	8.61	7	100	E	[53]
8-Methoxy[G-^{3}H]psoralen	$CDCl_3$	6.31 7.31 7.64	3 5 7	79.8 8.9 11.3	H	[23]

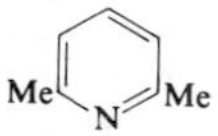

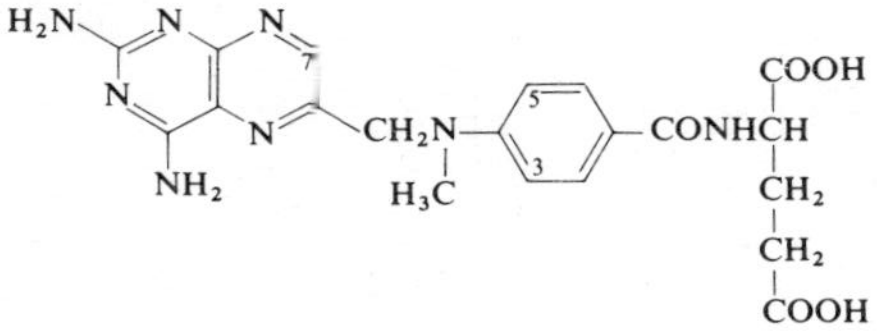

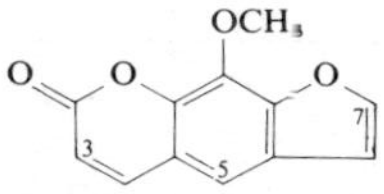

Table 30. (*contd.*)

Compound	Solvent	Chemical shift δ (p.p.m.)	Assignment	Distribution of tritium (%)	Method of preparation	Reference
[G-^{3}H]7-Methylbenz[c]acridine	$CDCl_3$	3.12 3.15	CHT_2 CH_2T	71.4	H	[23]
		7.70 7.81 7.86 7.91 8.08	Ring-^{3}H	6.3 8.7 9.5 3.2 0.8 (28.6)		
[G-^{3}H]12-Methylbenz[a]acridine	$CDCl_3$	3.64	Methyl	24.3	H	[23]
		7.89 7.93 8.13 8.25 8.9	Ring-^{3}H	60.1 10.1 2.3 3.2 (75.7)		
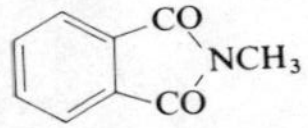 *N*-[^{3}H]Methylphthalimide	d_6-DMSO	3.03	Methyl	100	A	[23]
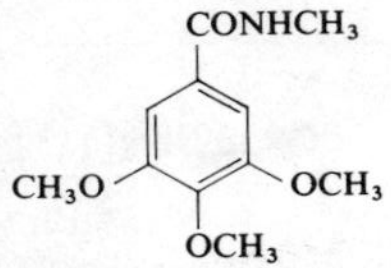 *N*-Methyl-3,4,5-trimethoxy-[*amino*-^{3}H]benzamide	d_6-DMSO	8.40	*N*-H	100	J	[113]
	for related amides δ_T(NT) was 7.58–9.09					

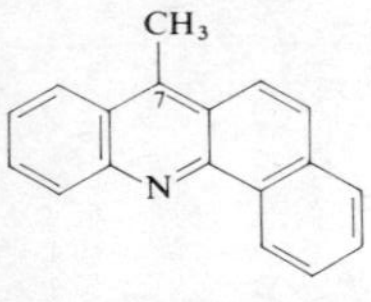

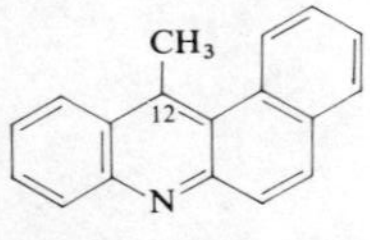

$Ph_3\overset{+}{P}Me$ I^-						
[^{3}H]Methyltriphenyl phosphonium iodide	d_6-DMSO	3.083d 3.103t 3.125q	*P*-CT_3 *P*-CHT_2 *P*-CH_2T	7 41 52	A	[19]
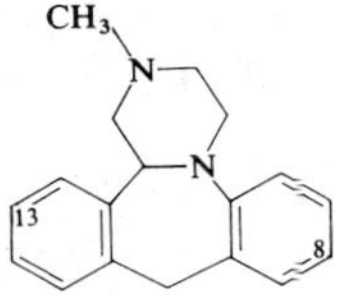	d_6-DMSO	6.88	8	100	D	[49]
[^{3}H]Mianserin	C_6D_6	6.94	13	100	D	[114]
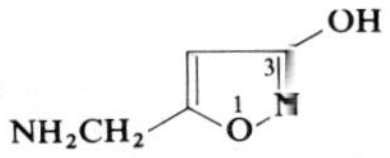						
[amino*methylene*-^{3}H]Muscimol	d_6-DMSO	3.56	*N*-CH_2	100	J	[49]
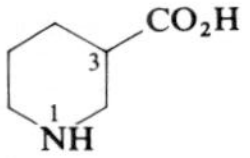						
DL-[3,4-^{3}H]Nipecotic acid	D_2O	2.56 1.73 1.99	3 ax. 4 ax. 4 eq.	36 6 58	C	[49]

Table 30. (*contd.*)

Compound	Solvent	Chemical shift δ (p.p.m.)	Assignment	Distribution of tritium (%)	Method of preparation	Reference
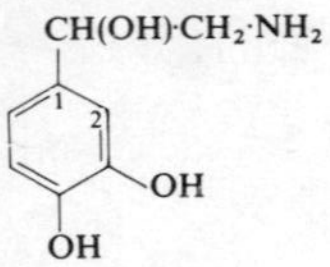	D_2O	3.2 4.8	N-CH_2 O-CH	63 37	C	[49]
[^{3}H]Norepinephrine	D_2O	4.82	O-CH	100	B	[49]
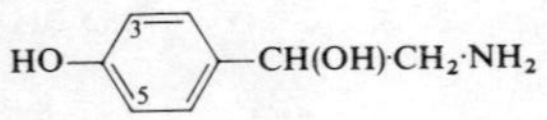 DL-[3,5-^{3}H]Octopamine hydrochloride	D_2O	6.98	Phenyl-3,5	100	D	[49]
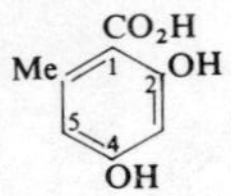 [3,5-^{3}H]Orsellinic acid	d_6-acetone	6.26 6.34	3 5	50 50	I	[115]
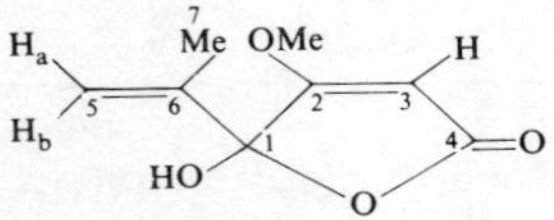 [^{3}H]Penicillic acid	d_6-acetone	5.27 s 5.49 m 5.19 m 1.77t	3 5a 5b 7	(a) (b) 18.5 58 2.2 12 10.1 30 69.2 —	K	[115,116]

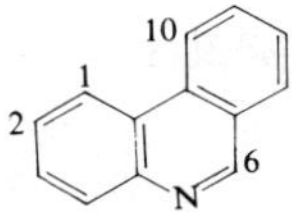 [G-^{3}H]Phenanthridine	18% in $CDCl_3$	8.43 7.67 7.56 8.17 9.17 7.90 7.70	1,10 2 3,8 4 6 7 9	2 14 34 11 10 13 16	G	[108]
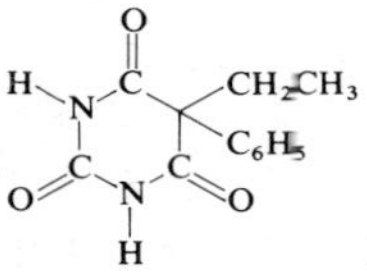 [^{3}H]Phenobarbitone	$CDCl_3$	7.43 7.37	Phenyl-3,5 Phenyl-4	60 40	H	[107]
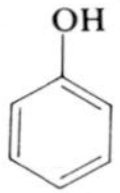 [^{3}H]Phenol	Neat	7.36 7.56 7.30	2,6 3,5 4	60.4 14.0 25.5	J	[94]
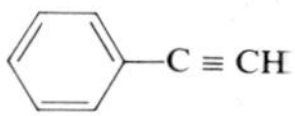 Phenyl[^{3}H]acetylene	Neat	3.05	Methine	100	I	[13]
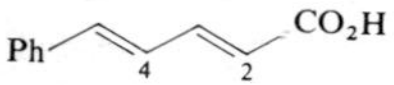 5-Phenyl[2,4-^{3}H]penta-2,4-dienoic acid	d_6-DMSO	6.08 s 7.13 s	2 4	74 26	K	[109]

Table 30. *(contd.)*

Compound	Solvent	Chemical shift δ (p.p.m.)	Assignment	Distribution of tritium (%)	Method of preparation	Reference
C_6H_5–$\overset{\beta}{C}H_2CH_2\overset{\alpha}{C}H_2OH$ 3-Phenyl[^{3}H]propanol	$CDCl_3$	3.1 1.4	α-CH_2 β-CH_2	96 4	J	[93]
Phthalic acid mono[^{3}H]methylamide ($C_6H_4(CO_2H)(CONHCH_3)$)	D_2O	2.87	*N*-methyl	100	A	[23]
	d_6-DMSO	2.71	*N*-methyl	100	A	[23]
2-[G-^{3}H]Picoline (2-methylpyridine)	Neat	2.45 7.07 7.47 7.02 8.55	Methyl 3 4 5 6	47 13 13 15 12	G	[108]
3-[G-^{3}H]Picoline (3-methylpyridine)	10% in CCl_4	2.27 8.37 7.40 7.10 8.35	Methyl 2 4 5 6	10 30 17 19 24	G	[108]

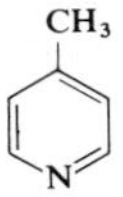 4-[G-^{3}H]Picoline	10% in CCl_4	2.29 8.38 7.03	Methyl 2,6 3,5	21 40 39	G	[108]
 [G-^{3}H] Piperidine hydrochloride	D_2O	3.10 1.57,1.61 1.69,1.72	2,6 3-,4- 5-CH_2	7 93	C	[49]
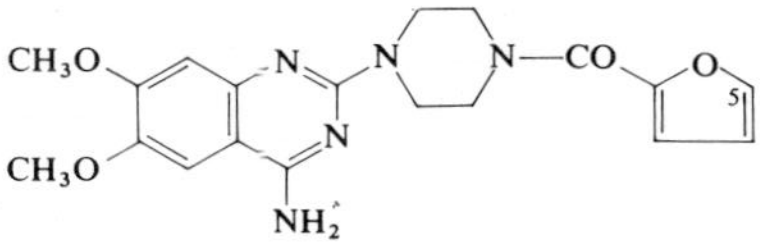 [^{3}H]Prazosin hydrochloride	d_6-DMSO	7.93	Furanyl-5	100	D	[49]
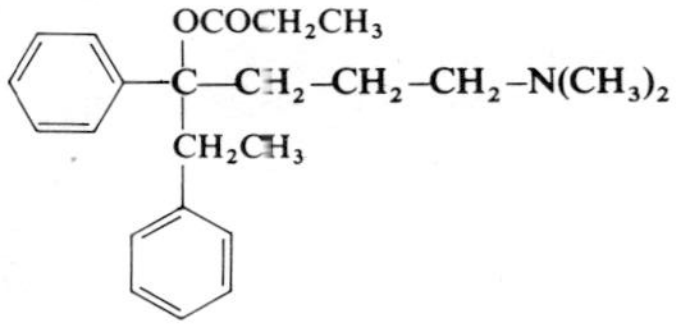 [^{3}H]Propoxyphene	D_2O	7.40 7.34	Aromatic	100	H	[107]

Table 30. (*contd.*)

Compound	Solvent	Chemical shift δ (p.p.m.)	Assignment	Distribution of tritium (%)	Method of preparation	Reference
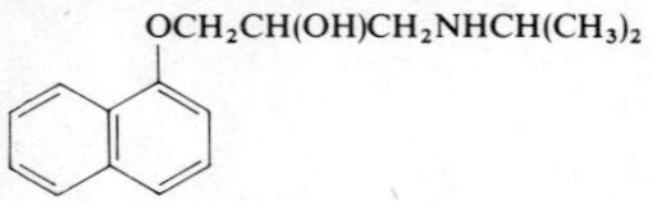 $OCH_2CH(OH)CH_2NHCH(CH_3)_2$ (1-naphthyl)	d_6-DMSO	7.54	Ring-4	100	D	[49]
[³H]Propranolol hydrochloride	d_6-DMSO	6.97	Ring-2	31	H	[107]
		7.43	Ring-3	10		
		7.54	Ring-4	13		
		7.50	Ring-6,7	46		
[G-³H]Pyridine	Neat	8.68	2,6	43	G	[108]
		7.27	3,5	40		
		7.66	4	17		
(structure: HOH_2C, CH_2OH, OH, CH_3 pyridine)	D_2O	2.63	Methyl	40	G	[23]
		5.0	4-CH_2OH	4.9		
		8.19	6	55.1		
[G-³H]Pyridoxine hydrochloride	D_2O	2.65	Methyl	63.4	E	[23]
		5.0	4-CH_2OH	22.4		
		4.82	5-CH_2OH	2.0		
		8.21	6	12.2		
(structure: 2-pyridyl-N(CH_2-C₆H₄-OMe)($CH_2CH_2NMe_2$)) [*pyridinyl*-5-³H]Pyrilamine	d_6-DMSO	6.65	Pyridinyl-5	100	D	[49]

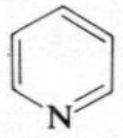

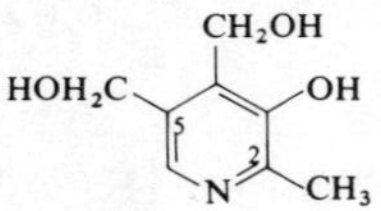

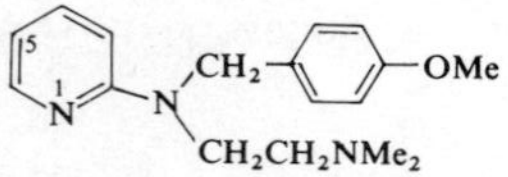

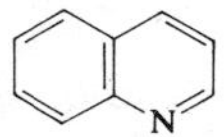		9.07	2	13		
		7.14	3	13		
		7.86	4	16		
		7.62	5	13	G	[108]
		7.39	6	14		
		7.64	7	19		
[G-^{3}H]Quincline	Neat	8.49	8	12		
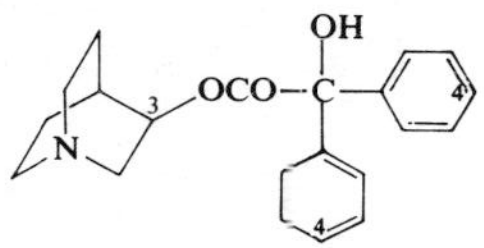						
DL-[3-^{3}H]Quinuclidinyl benzilate	d_6-DMSO	4.86	Quinuclidinyl-3	100	B	[49]
L-Quinuclidinyl [*phenyl*-4-^{3}H]-benzilate	D_2O	7.30	Phenyl-4	100	D	[49]
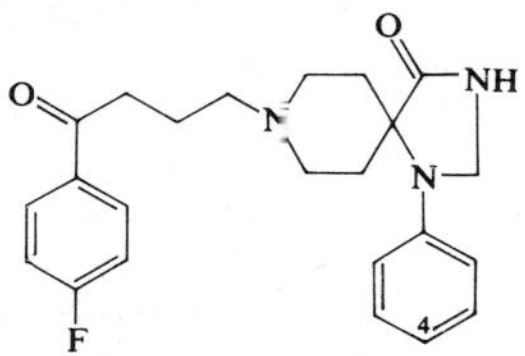						
[*phenyl*-4-^{3}H]Spiperone	d_6-DMSO	6.77	Phenyl-4	100	D	[49]

Table 30. *(contd.)*

Compound	Solvent	Chemical shift δ (p.p.m.)	Assignment	Distribution of tritium (%)	Method of preparation	Reference
CH_3CH_2CON (succinimide) *N*-Succinimidyl [2,3-^{3}H]propionate	$CDCl_3$	2.55 1.2	2 3	32.2 67.8	C	[23]
[N-*methyl*-^{3}H]Tamoxifen	$CDCl_3$	2.61	*N*-methyl	100	A	[23]
$Ph_4P^+Br^-$ Tetra[^{3}H]phenylphosphonium bromide	d_6-DMSO	8.016d [$^5J_{PT}$ = 2.2 Hz)	Phenyl	100	D	[23]

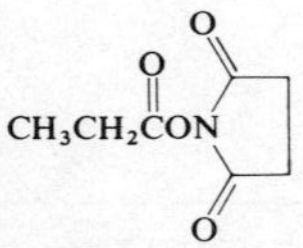

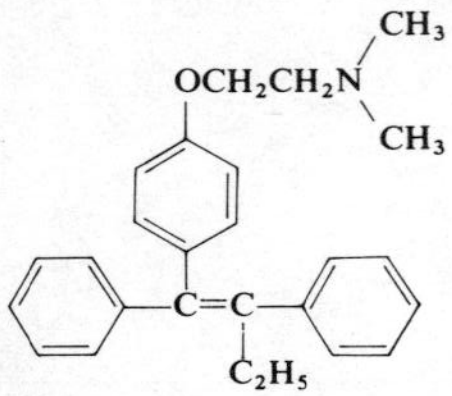

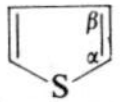 [^{3}H]Thiophen	Neat	7.00 6.88	α-CH β-CH	56 44	J	[94]
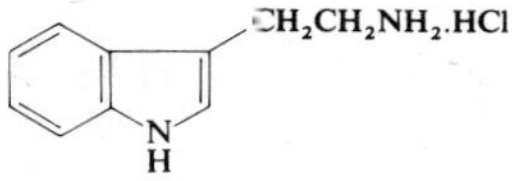 [G-^{3}H]Tryptamine hydrochloride	D_2O	7.34 7.24 7.59	2 Benzene ring Benzene ring	56.5 18.5 24.9	G	[23]
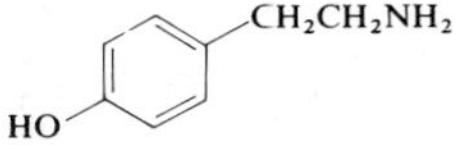 [*side-chain*-2-^{3}H]Tyramine hydrochloride	D_2O	2.88	Side chain-2	100	E	[23]
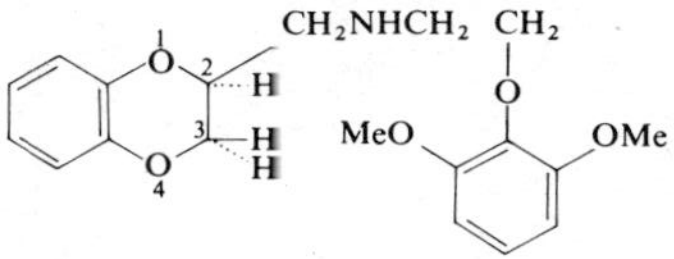 [dimethoxybenzene-*ring*-^{3}*H*]WB4101	d_6-DMSO	6.68	Dimethoxy ring	100	D	[23]
[2,3-^{3}H]WB4101	d_6-DMSO	5.34 5.45 5.43	2α 3α 3β	(a) (b) 27 23 45 50 28 27	C	[116a]

ment in the 6-bromo derivative, did not result in specific labelling. The ^1H-decoupled ^3H nmr spectrum (Figure 32) shows two lines. Comparison with the ^1H nmr spectrum shows that the extra signal at $\delta = 9.01$ comes from the 1-position, which is apparently sufficiently acidic to exchange with the tritium under the hydrogenation conditions.

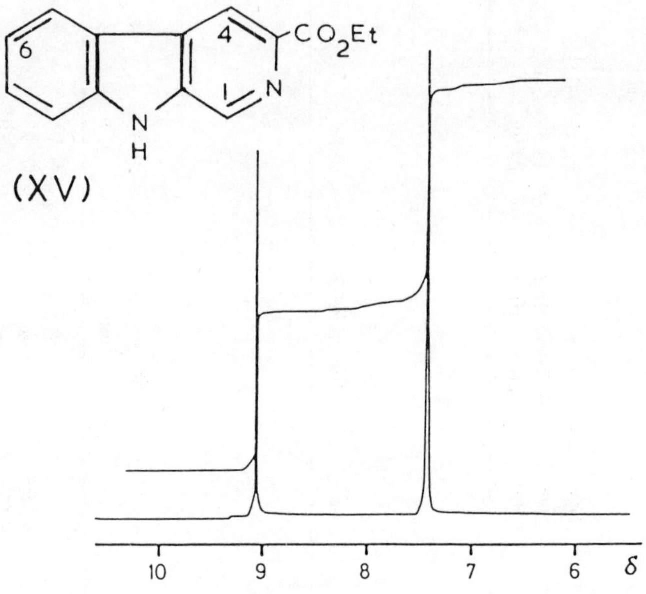

Figure 32. ^3H Nmr spectrum (^1H-decoupled) of ethyl [^3H]β-carboline-3-carboxylate in d_6-DMSO. (Reproduced by permission of D. Reidel Publishing Company.)

Reduction of a carboxyl-protected 4-carboxypyridine with sodium borotritide yielded tritiated isoguvacine (**XVI**) which shows two singlets in the ^1H-decoupled ^3H nmr spectrum (Figure 33a) at $\delta = 3.38$ and 3.87. The ^1H nmr spectrum of isoguvacine (Figure 33b) could be assigned unambiguously from the splitting pattern, so making the interpretation of the ^3H nmr spectrum immediately obvious [54]. This result corrected earlier data for an impure product [49].

The enhanced, coupled, ^3H nmr spectrum of [^3H]methyltriphenylphosphonium iodide (Figure 34), prepared by methylation of triphenylphosphine with very high specific activity [^3H]methyl iodide, shows fully resolved, but interleaved, doublet, triplet and quartet signals from the PCT_3, PCT_2H and $PCTH_2$ groups present, the coupling constants $^2J_{TP}$ and $^2J_{TH}$ happening to be equal [19].

The preparation of *N*-acetyl-5-hydroxy[^3H]tryptamine (*N*-acetyl[^3H]serotonin) by catalytic replacement of bromine for tritium in *N*-acetyl-4,6-dibromoserotonin gave the tritiated compound at much higher specific activity than

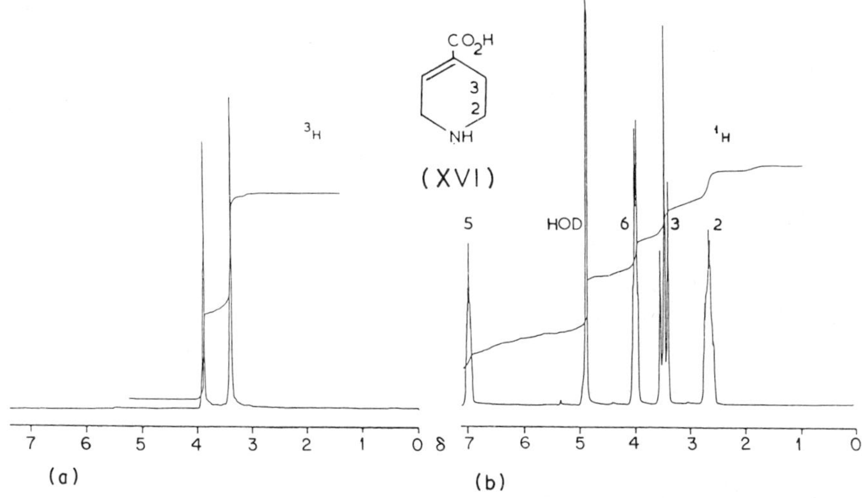

Figure 33. Nmr spectra of [^{3}H]isoguvacine in D_2O: (a) ^{3}H spectrum acquired with ^{1}H-decoupling and (b) ^{1}H spectrum

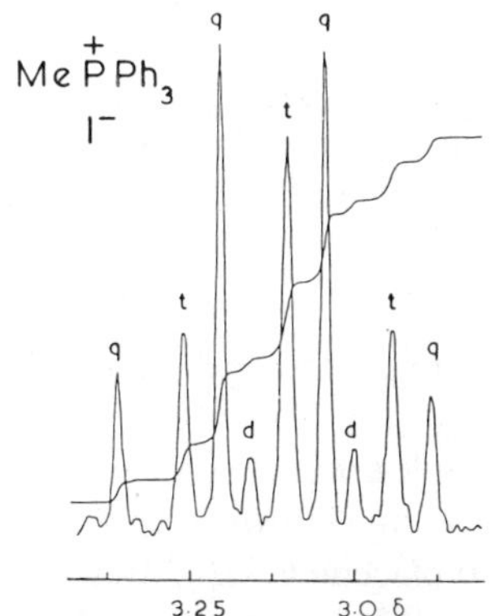

Figure 34. Resolution enhanced, coupled, ^{3}H nmr spectrum of [*methyl*-^{3}H]methyltriphenylphosphonium iodide in d_6-DMSO. Fortuitous equal coupling to phosphorus and protons gives rise to the quartet, triplet and doublet signals from the PCH_2T, $PCHT_2$ and PCT_3 groups present. (Reproduced by permission of John Wiley & Sons Ltd.)

expected, suggesting that labelling occurs elsewhere than just in the 4,6-positions. Indeed, the ^{1}H-decoupled ^{3}H nmr spectrum (Figure 35a) showed three lines at $\delta = 6.63$, 6.86 and 7.07, whilst the ^{1}H-coupled ^{3}H nmr spectrum (Figure 35b) showed that the signal at $\delta = 6.63$ was a double doublet ($J_{TH} = 9$, 2.2 Hz) and therefore assignable to the 6-position. By adding D_2O to exchange the 1-NH proton and so remove the coupling to the 2-position, the signal at $\delta = 7.07$ sharpened to a singlet (Figure 35c) and could then be assigned to the 2-position

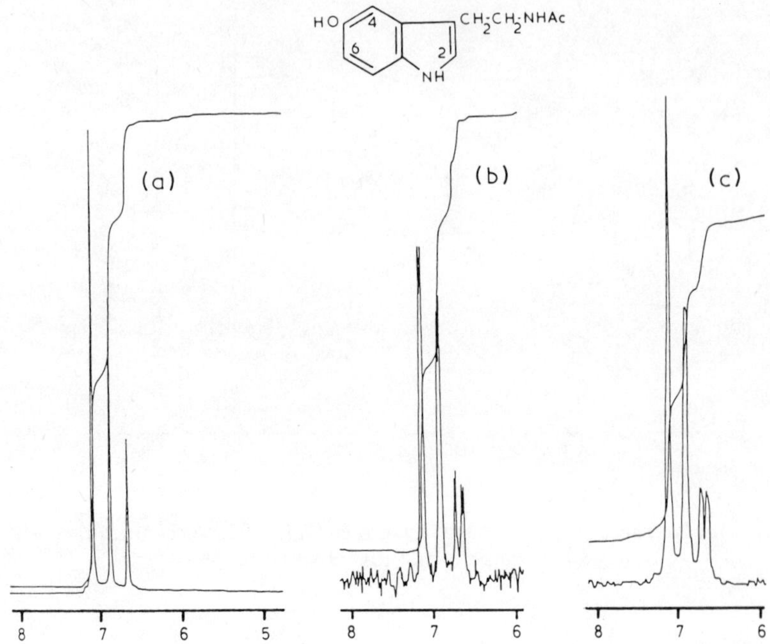

Figure 35. ^{3}H Nmr spectra of *N*-acetyl[^{3}H]serotonin in d_6-DMSO: (a) ^{1}H-decoupled, (b) ^{1}H-coupled and (c) as in (b) after addition of a trace of D_2O. (Reproduced by permission of D. Reidel Publishing Company.)

with confidence. Hence the signal at $\delta = 6.86$ was assigned to the 4-position, the meta-coupling to 6-H being just discernable in the ^{1}H-coupled ^{3}H nmr spectra (Figure 35b and c). This example provides another instance of an acidic proton (at the 2-position, adjacent to a ring nitrogen atom) being exchanged under the catalysed halogen–tritium replacement conditions.

A compound which gave a ^{3}H nmr spectrum of some interest was 4-fluoro-3-nitro[2,6-^{3}H]phenylazide ([2,6-^{3}H]FNPA). The labelling had been introduced by catalytic tritio-debromination of 2,6-dibromo-4-fluorobenzenamine; the resultant intermediate was then nitrated, diazotized and treated with sodium azide. Entirely specific, almost equal labelling was thus effected in the [2,6-^{3}H]FNPA, as was shown by the two signals at about $\delta = 7.30$ and 7.75 in the ^{3}H nmr spectrum acquired with ^{1}H decoupling (see Figure 36). The higher field signal, from the 6-triton in the doubly tritiated molecules, comprised a 1:2:1 triplet at $\delta = 7.306$, which therefore was indicative of equal tritium–tritium and tritium–fluorine coupling constants such that $^3J\,(2T, 6T) = {}^4J\,(4F, 6T) = 3.4$ Hz. In contrast, the lower field signal from the 2-triton appeared as a double doublet, at $\delta = 7.753$, giving $^3J\,(2T, 6T) = 3.4$ Hz and $^4J\,(4F, 2T) = 6.4$ Hz. Thus the 4,2-meta fluorine–tritium coupling constant had nearly twice the magnitude of the 4,6-meta coupling constant in [2,6-$^3H_2$]FNPA.

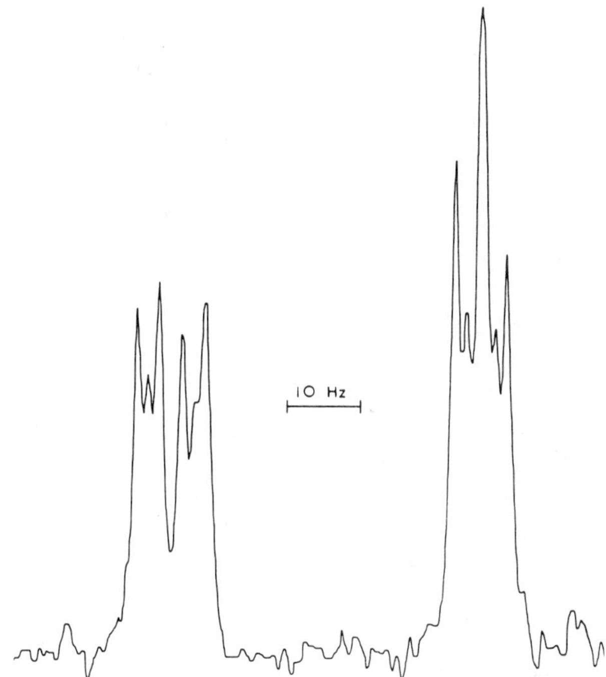

Figure 36. ^3H Nmr spectrum of 4-fluoro-3-nitro[2,6-^3H]phenylazide (FNPA) in $CDCl_3$, acquired with ^1H-decoupling

A further feature of this ^1H-decoupled ^3H nmr spectrum (Figure 36) comprised two doublet signals, respectively interleaved in the two preceding multiplets. That doublet in the higher field signal was centred at $\delta = 7.307$ and arose from a 6-triton coupled only to 4-fluorine with $J = 3.4$ Hz, indicating the presence of singly labelled 6-tritiated molecules. The doublet in the lower field signal, centred at $\delta = 7.754$ had $J = 6.4$ Hz, from a 2-triton coupled only to 4-fluorine. This was indicative of the presence of singly labelled 2-tritiated species.

Yet another observation from the spectrum (Figure 36) was the shift to a higher field of each tritium resonance in the doubly labelled species. This isotopic shift amounted to 0.001 p.p.m., exerted over four bonds (cf. pages 11 and 209).

(viii) Tritium labelled alkaloids and related compounds

The alkaloids, which can be described as basic compounds with marked physiological action, occur largely in the dicotyledonous plants. These compounds comprise an enormous range of chemical types, are usually classified on the basis of their nitrogenous function and over the years their study has generated much important organic chemistry. In recent times numerous tritiated alkaloids have found uses as radioactive ligands in studies of the neurotransmitters of the central nervous system. Such studies have been aimed at achieving a better understanding and possible control of disorders of the nervous system, e.g. in Huntington's chorea, Parkinsonism, depressive illness and schizophrenia. Examples of tritiated alkaloids used for the study of neurochemical receptors [117] are atropine (muscarinic receptor), dihydromorphine, diprenorphine and etorphine (opiate receptors), 9,10-dihydroergocryptine (α-adrenergic receptors), lysergic acid diethylamide (serotonin receptors) and strychnine sulphate (glycine receptors). Table 31 lists the chemical shifts and distribution of labelling for these tritiated alkaloids and related compounds, as found from their ^{1}H-decoupled ^{3}H nmr spectra.

One of the most interesting set of data is that obtained for [^{3}H]atropine, prepared by exposure to tritium gas (Wilzbach [56]). The ^{1}H-decoupled ^{3}H nmr spectrum (Figure 37a) shows that most of the label is in the *N*-methyl group—quite an unexpected result from this type of labelling procedure. There is evidence for both $-CH_2T$ and $-CHT_2$ species and this is confirmed in the coupled spectrum (Figure 37b) where there is a triplet at $\delta = 2.23$ with J(gem. HT) $= 12.8$ Hz from the $N–CH_2T$ group and a doublet at $\delta = 2.20$ with the same splitting from the $N–CHT_2$ group. In addition there is a singlet at $\delta = 2.18$ from the $N–CT_3$ group; this signal is obscured in the decoupled spectrum. The chemical shifts of the non-equivalent hydrogens of the acyl–CH_2 group are available by inspection of the ^{1}H-decoupled spectrum, as is the shift of the acyl–CH group. These last three assignments are confirmed in the coupled spectrum, where each of the three signals appears as the four-line X part of an ABX spin-coupled system, with two of the four lines from the CTPh signal overlapping the two lowest field lines of the adjacent –CHT–O– group multiplet. Figure 37c shows the detailed interpretation.

Lysergic acid diethylamide (LSD) on bromination and subsequent catalysed halogen–tritium replacement gave [2-^{3}H]LSD. This completely specific bromination and tritiation was confirmed by the ^{3}H nmr spectrum (Figure 38) which shows a single line at $\delta = 7.22$ indicative of 2-tritiation. Equally selective was the tritiation of a dehydrodiprenorphine with sodium borotritide to give specifically [16-^{3}H]diprenorphine, as demonstrated by the single sharp ^{3}H resonance at $\delta = 2.37$ in the ^{1}H-decoupled ^{3}H nmr spectrum.

The ^{3}H nmr spectrum of [G-^{3}H]vinblastine sulphate (**XVII**) (Figure 39) illustrates the extreme convenience of ^{3}H nmr spectroscopy for the examination of products with complex molecules which have been tritiated by a general labelling procedure [62]. Vinblastine, which is used for the effective treatment of certain cancers, is a light-sensitive compound which cannot be subjected to vigorous experimental conditions. Nevertheless, the drug can be tritiated by treatment (as the sulphate) with [*carboxyl*-^{3}H]trifluoracetic acid (CF_3COOT) at room temperature (20 °C) for 2 hours. The ^{3}H nmr spectrum showed that the tritium was confined to stable aromatic positions with 44 per cent. at the 17-position and the rest in the indole 6-membered ring, with 46 per cent. at the 12′, 13′-positions and 10 per cent. distributed between the 11′- and 14′-positions. The assignments in the indole ring were made partly from the expectation that the

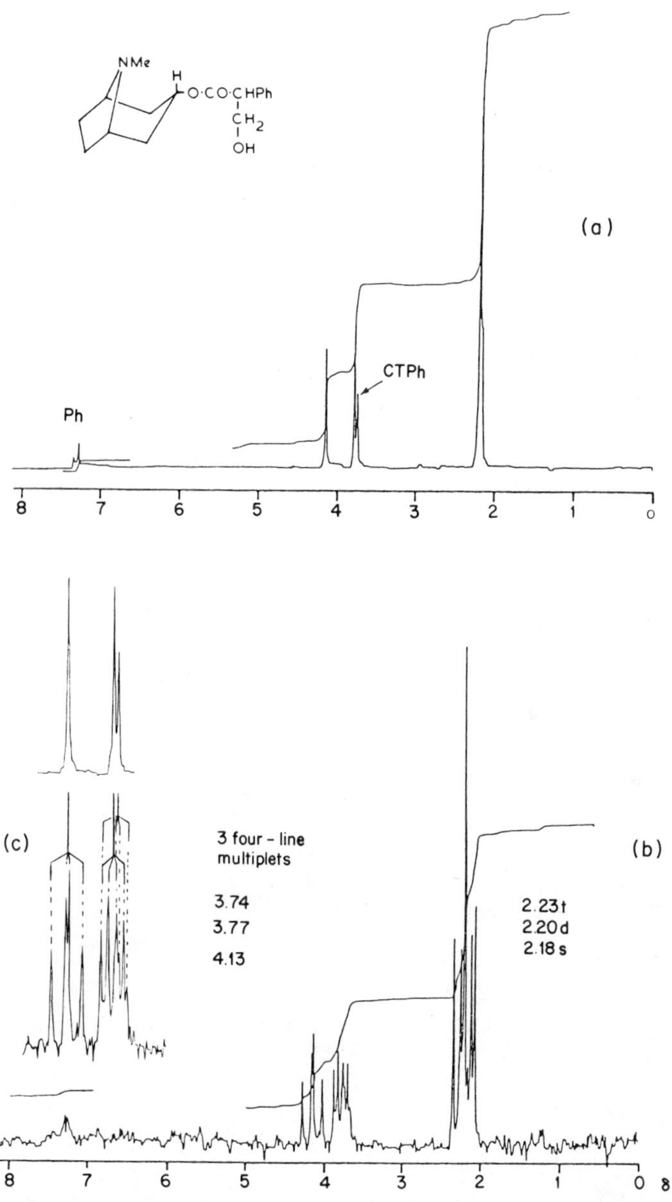

Figure 37. ^{3}H Nmr spectra of [G-^{3}H]atropine: (a) ^{1}H-decoupled, (b) ^{1}H-coupled and (c) the splitting diagram for the acyl CH_2 and CH signals. ((a), (b) reproduced by permission of The Royal Society of Chemistry.)

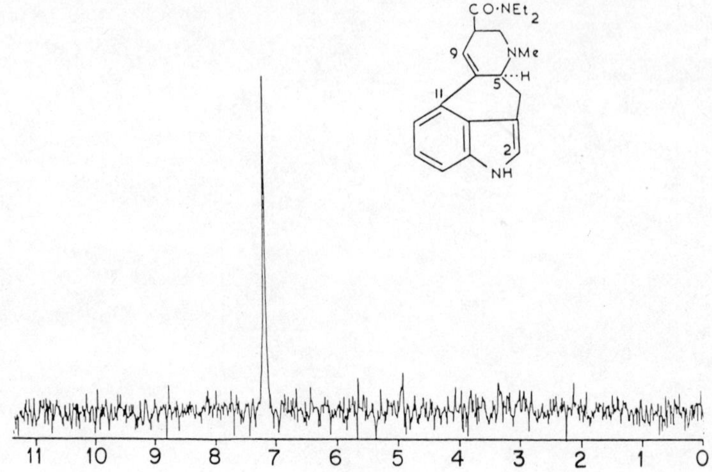

Figure 38. ^{3}H Nmr spectrum of [^{3}H]lysergic acid diethylamide in d_6-DMSO, acquired with ^{1}H-decoupling. (Reproduced by permission of D. Reidel Publishing Company.)

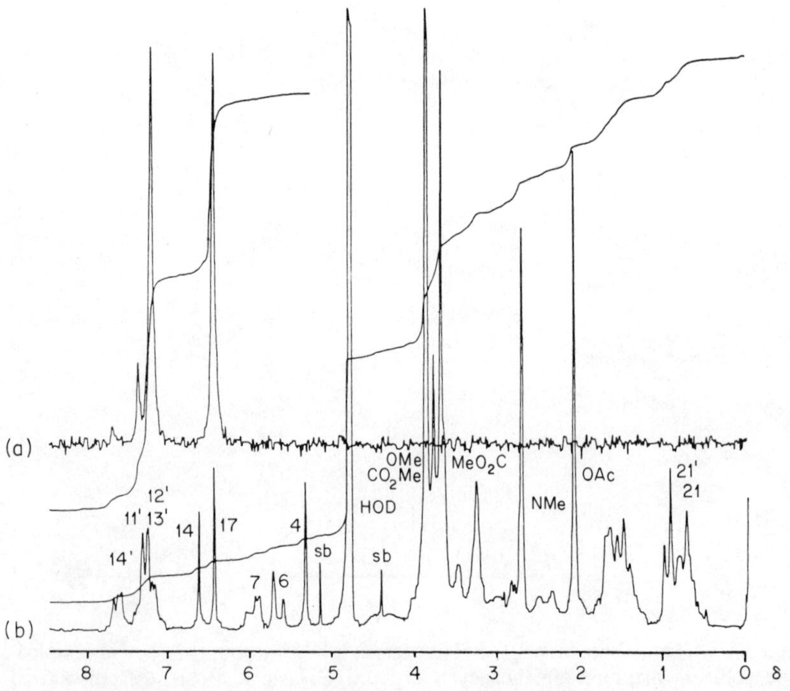

Figure 39. (a) ^{3}H Nmr spectrum of [G-^{3}H]vinblastine sulphate in D_2O, acquired with ^{1}H-decoupling, superimposed on (b) the ^{1}H spectrum. (Reproduced by permission of The Royal Society of Chemistry.)

Table 31. Chemical shifts and distribution of labelling for some tritium labelled alkaloids and related compounds.

Compound	Solvent	Chemical shift δ (p.p.m.)	Assignment	Distribution of tritium (%)	Method of preparation	Reference
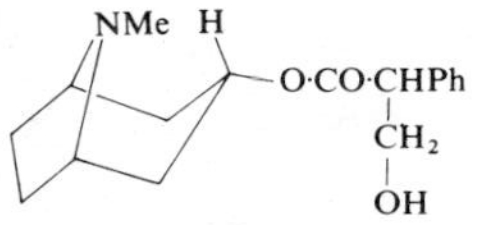 [G-^{3}H]Atropine	$CDCl_3$	2.18, 2.20, 2.23	*N*–Me (CHT_2, CH_2T, CT_3)	59	F	[49, 104]
		3.74	Benzylic CH	6		
		3.77, 4.13	*O*–CH_2	30		
		7.29, 7.35	Phenyl	5		
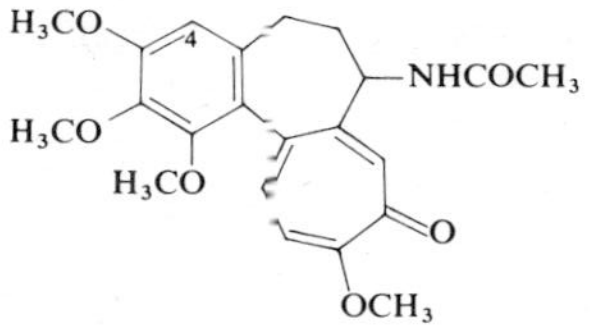 [*ring* A-4-^{3}H]Colchicine	D_2O	6.76	4	100	H	[23]
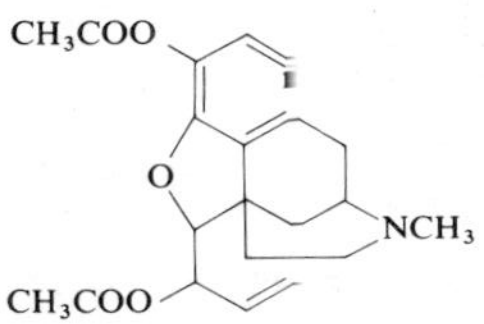 Diacetyl[1-^{3}H]morphine	d_6-DMSO	6.77	1	100	D	[49]

Table 31. *(contd.)*

Compound	Solvent	Chemical shift δ (p.p.m.)	Assignment	Distribution of tritium (%)	Method of preparation	Reference
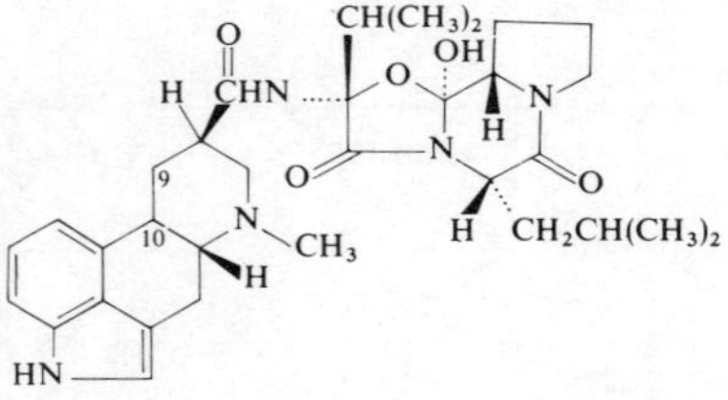		2.93	8	4		
		2.47	9	13		
		2.61		37		
		2.75	10	43	C	[49]
		3.35	CH.CO of dipeptide	3		
[9,10(*n*)-^{3}H]-9,10-Dihydroergocryptine	d_6-DMSO					
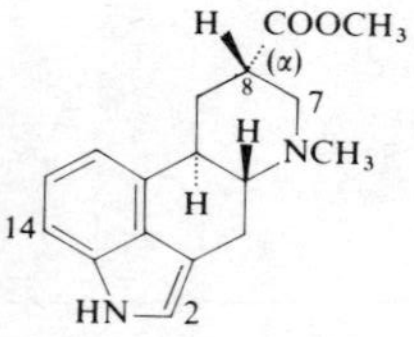		7.0	2	42		
		2.3	7 ax.	7		
	d_6-DMSO	2.8	8	26	C	[55]
[G-^{3}H]-9,10-Dihydroisolysergic acid methyl ester		7.2	14	25		

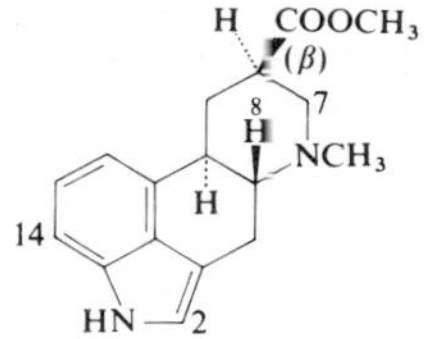 [G-^{3}H]-9,10-Dihydrolysergic acid methyl ester	d_6-DMSO	7.0 2.8 7.2	2 9 eq., 10ax. 14	2 97 0.5	C	[55]
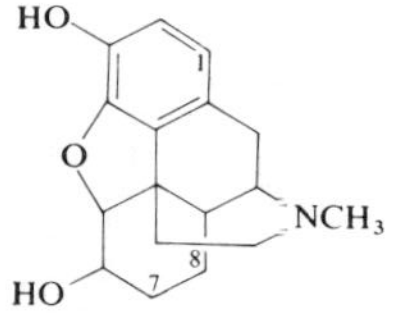 [1,7,8-^{3}H]Dihydromorphine	d_6-DMSO	6.67 1.35	1 7, 8	36 64	C, D	[49]
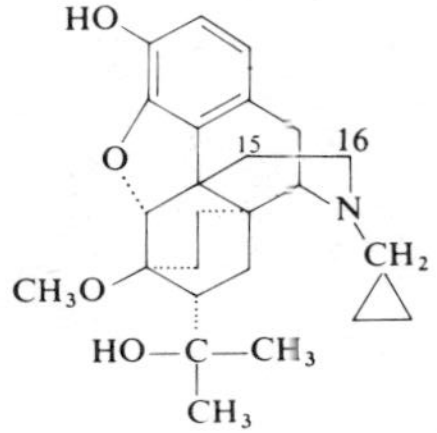 [^{3}H]Diprenorphine	$CDCl_3$	(a) 2.62	16	100	B	[23]
		(b) 2.37	16	100		[49]
	$CDCl_3$	(c) 1.89 2.78	15 16	40 60	C	[23]
		(d) 1.83 2.78	15 16	37.5 62.5	C	[23]
		(a) + (d) 1.88 2.78	15 16	27.1 72.9		[23]

Table 31. (*contd.*)

Compound	Solvent	Chemical shift δ (p.p.m.)	Assignment	Distribution of tritium (%)	Method of preparation	Reference
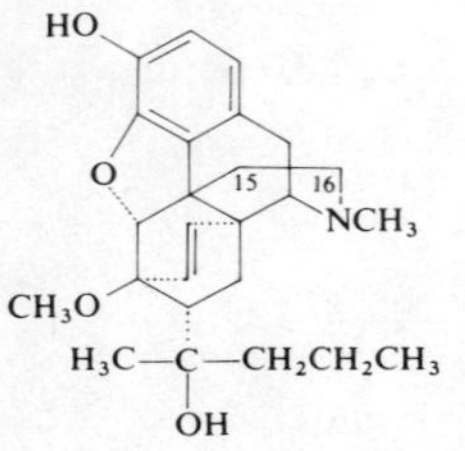 [15,16-^{3}H]Etorphine	d_6-DMSO	1.85 2.75	15 16	43 57	C	[49]
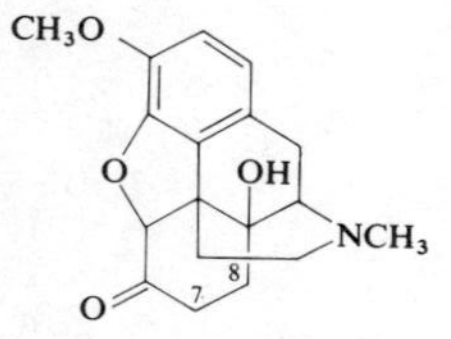 14-Hydroxy[7,8-^{3}H]dihydrocodeinone	d_6-DMSO	2.85–3.02 2.07 1.92 1.47	7α 7β 8α 8β	9.9 9.9 18.6 61.5	C	[23]
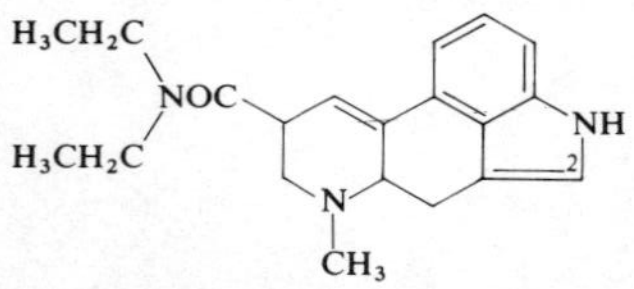 [2-^{3}H]Lysergic acid diethylamide	d_6-DMSO	7.22	2	100	D	[49]

Compound	Solvent	δ (ppm)	Position	%		Ref.
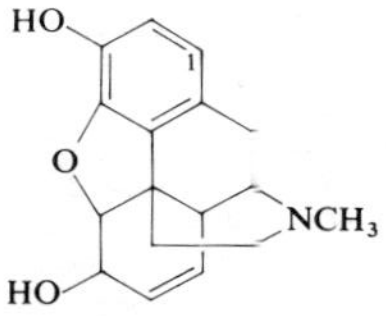 [1-^{3}H]Morphine	d_6-DMSO	6.59	1	100	D	[49]
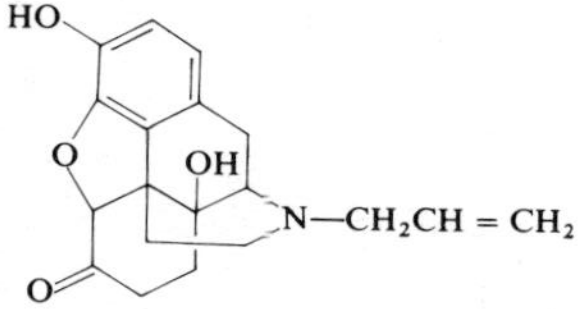 [*N-allyl*-2,3-^{3}H]Naloxone	d_6-DMSO	5.97	Allyl-2	50	C	[23]
		5.32	Allyl-3	50		
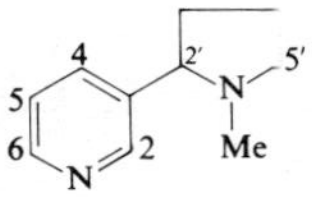 [G-^{3}H]Nicotine	Neat	8.68	2	4.9	G	[108]
		7.72	4	1		
		7.25	5	18.3		
		8.54	6	19.3		
		3.02	2′	1		
		1.65	3′a	11.9		
		2.19	3′b	15.9		
		1.81	4′a	10.9		
		2.08	4′b	6.9		
		3.12	5′b	9.9		

Table 31. (*contd.*)

Compound	Solvent	Chemical shift δ (p.p.m.)	Assignment	Distribution of tritium (%)	Method of preparation	Reference
[G-^{3}H]Nicotine *d*-bitartrate	D_2O	9.02	2	11.6	G	[23]
		8.73	4	8.3		
		8.15	5	12.0		
		8.92	6	11.0		
		3.40	2′	12.4		
		2.30	3′a	5.1		
		2.67	3′b	3.2		
		2.32	4′a	6.5		
		2.41	4′b	1.9		
		2.83	5′a	18.8		
		3.89	5′b	9.2		
HO, O, OH, NCH_3, O, 1, 2, 7, 8 (structure) [1,2,7,8-^{3}H]Oxymorphone	d_6-DMSO	6.71	1, 2	47.9	C, D	[23]
		1.89	7	39.8		
		1.43, 1.51	8	12.3		
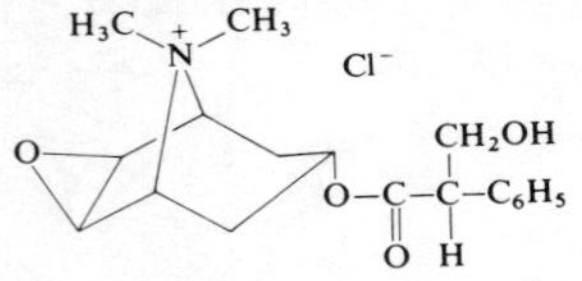 H_3C, $\overset{+}{N}$, CH_3, Cl^-, O, $O-\underset{\underset{O}{\parallel}}{C}-\underset{H}{\overset{CH_2OH}{C}}-C_6H_5$ (structure) [*N-methyl*-^{3}H]Scopolamine methochloride	D_2O	3.242	CT_3 } *N*-methyl	100	A	[23]
		3.271	CT_2H			
		3.297	CTH_2			

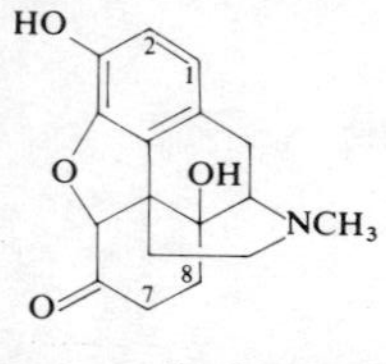

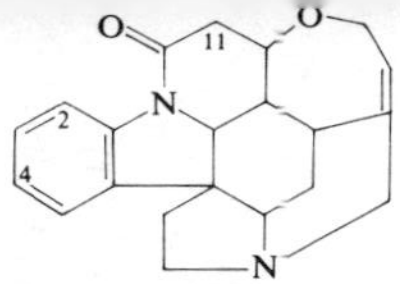

Compound	Solvent	δ	Position	%		Ref.
[2,4,11-^{3}H]Strychnine sulphate	$CDCl_3$	7.20	2	24	H	[49]
		8.12	4	23		
		3.15	11α	26		
		2.65	11β	27		
[G-^{3}H]Vinblastine sulphate	D_2O	7.40	11′	8	H	[62]
		7.25	12′	} 46		
		7.25	13′			
		7.72	14′	2		
		6.45	17	44		
[G-^{3}H]Vincristine sulphate	D_2O	7.40	11′	12.2	H	[54]
		7.25	12′	} 81.7		
		7.25	13′			
		7.73	14′	6.1		

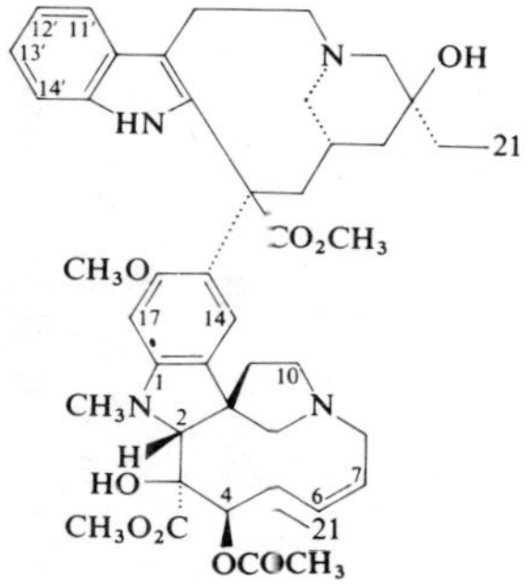

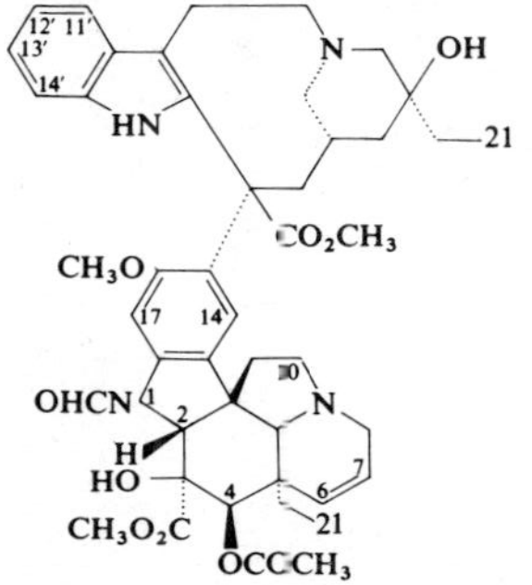

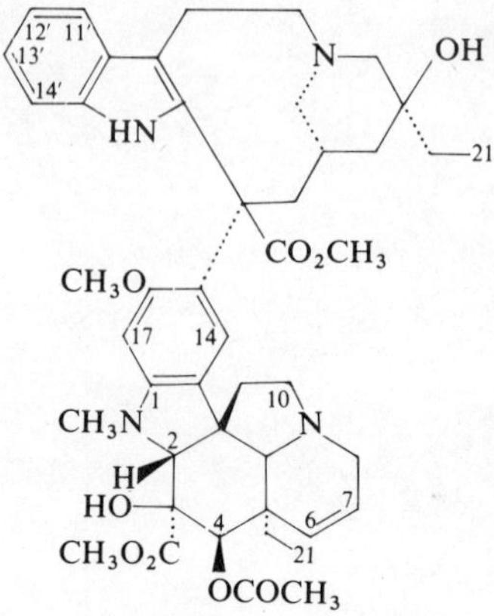

(XVII)

signal from H-14′ would be relatively broadened by the adjacent quadrupolar nitrogen. It would surely have been an extremely difficult or even impossible task in this case to locate the tritium labelled positions quantitatively by conventional chemical degradation procedures.

The ^3H nmr assignments for [^3H]strychnine (**XVIII**) were made in part by comparison of the ^1H nmr spectrum (at 250 MHz) with the ^1H-decoupled ^3H nmr lines [49]. Useful confirmation of assignments came partly from the ^1H-coupled ^3H nmr spectrum, using the signal multiplicities, and partly from a specific ^1H-decoupling experiment. Because the signal at $\delta = 7.21$ in the ^1H-coupled ^3H nmr spectrum was a triplet, it was necessarily assignable to either the 2- or the 3-position of the strychnine molecule (two ortho-H coupled to T). Decoupling of 4-^1H at $\delta = 8.1$ in the proton region failed to affect the ^3H triplet in the ^3H region, indicating that the triton responsible could not be in the adjacent 3-position: hence the triton was assigned to the 2-position [49].

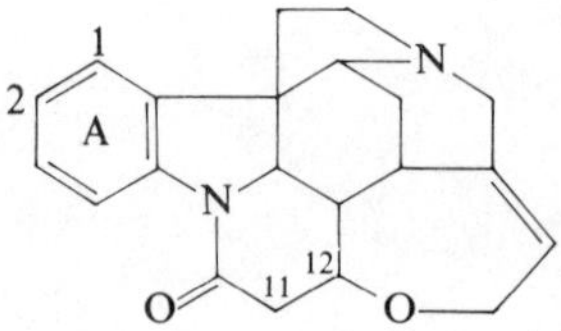

(XVIII)

(ix) Tritium labelled aliphatic and alicyclic compounds

Although most attention has been focused on the ^3H nmr spectra of tritiated amino acids, steroids and aromatic and heterocyclic compounds, the final listing in this chapter, a selection of aliphatic and alicyclic compounds, is no less

Table 32. Chemical shifts and distribution of labelling for some tritium labelled aliphatic and alicyclic compounds

Compound	Solvent	Chemical shifts δ (p.p.m.)	Assignment	Distribution of tritium (%)	Method of preparation	References
cis-3-Amino[^{3}H]cyclohexane carboxylic acid	D_2O	1.33 1.37	Some or all of 4 ax., 5 ax., 6 ax.	25 18	C	[49]
		1.88 1.96 1.99	4 eq., 5 eq., 6 eq.	17 15 25		
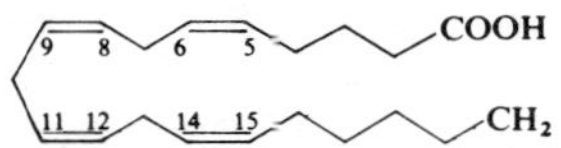 [5,6,8,9,11,12,14,15-^{3}H]Arachidonic acid	$CDCl_3$	5.38	5,6,8,9,11,12,14,15	100	C	[23]
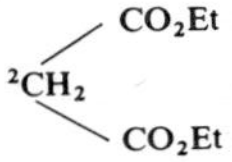 Diethyl [2-^{3}H]malonate	d_6-acetone	3.25	2	100	I	[115]
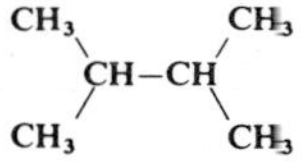 2,3-[G-^{3}H]Dimethylbutane	Neat	1.27 0.84	CH CH_3	17 83	J	[68]

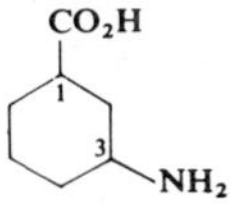

Table 32. (*contd.*)

Compound	Solvent	Chemical shifts δ (p.p.m.)	Assignment	Distribution of tritium (%)	Method of preparation	References
[8,9,11,12,14,15-^{3}H]Eicosa-8,11,14-trienoic acid	$CDCl_3$	5.25	8,9,11,12,14,15	100	C	[23]
$CH_3 . CH_2OH$ [1-^{3}H]Ethanol	D_2O	3.1	CH_2	100	J	[93]
$H_2N \cdot CH_2CH_2OH$ [1-^{3}H]Ethanolamine hydrochloride	D_2O	3.76	O-CH_2	100	B	[23]
CH_2OH | $CHOH$ | CH_2OH [2-^{3}H]Glycerol	D_2O	3.74	2	100	B	[23]
$CH_2O.CO(CH_2)_7CH{=}CH(CH_2)_7CH_3$ | $CHO.CO(CH_2)_7CH{=}CH(CH_2)_7CH_3$ | $CH_2O.CO(CH_2)_7CH{=}CH(CH_2)_7CH_3$ [2-^{3}H]Glycerol trioleate	$CDCl_3$	5.25	2	100	B	[23]
$CH_3(CH_2)_4\overset{\beta}{C}H_2\overset{\alpha}{C}H_2OH$ [^{3}H]Heptanol	$CDCl_3$	3.6 1.5	α-CH_2 β-CH_2	(a) (b) 100 86 — 14	J	[93]

$CH_3(CH_2)_9\overset{\alpha}{C}H_2CO_2H$ [^{3}H]Lauric acid	$CDCl_3$	2.11	α-CH_2	100	I	[65]
CH_3NH_2 [^{3}H]Methylamine hydrochloride	D_2O	2.57	Methyl	100	A	[23]
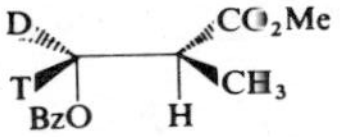 (structure: D, T, BzO / CO_2Me, CH_3, H) Methyl 3-benzoyloxy[3-*pro S*-^{3}H]isobutyrate	$CDCl_3$	5.80	3-*Pro S*	100	C, K	[37, 118]
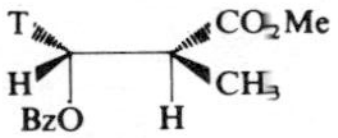 (structure: T, H, BzO / CO_2Me, CH_3, H) Methyl 3-benzoyloxy[3-*pro R*-^{3}H]isobutyrate	$CDCl_3$	5.96	3-*Pro R*	100	C, K	[37, 118]
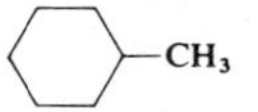 (cyclohexane–CH_3) [^{3}H]Methylcyclohexane	Neat	1.63 1.26 0.84	CH CH_2 CH_3	28 32 40	J	[68]
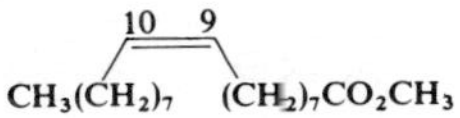 $CH_3(CH_2)_7\overset{10}{C}H{=}\overset{9}{C}H(CH_2)_7CO_2CH_3$ Methyl (*Z*)-[9,10(*n*)-^{3}H]octadec-9-enoate	d_6-benzene	5.427 } 5.441 } 5.432	9-^{3}H or 10-^{3}H 9,10-$^3H_2$	89–96	C	[119]
$CH_3CH_2CH(CH_3)CH_2CH_3$ 3-[^{3}H]Methylpentane	Neat	1.28 1.10 0.84	CH CH_2 CH_3	21 29 50	J	[68]

Table 32. (*contd.*)

Compound	Solvent	Chemical shifts δ (p.p.m.)	Assignment	Distribution of tritium (%)	Method of preparation	References
HO, CH_3 (structure); DL-[2-^{3}H]Mevalonic acid lactone	d_6-benzene	1.96 s 2.33 s	2 2	50 50	K	[23]
$CH_3(CH_2)_{11}\overset{\alpha}{C}H_2CO_2H$ [^{3}H]Myristic acid	$CDCl_3$	2.07	α-CH_2	100	I	[65]
$CH_3(CH_2)_4$ (4, 3) $\overset{+}{P}H_3I^-$ (structure) [3,4-^{3}H]Non-3-enyl-1-phosphonium iodide	CD_3OD	5.53 5.48	3 4	50 50	C	[23]
$CH_3(CH_2)_7$ (10, 9) $(CH_2)_7CH_2OH$ (structure) (*Z*)-[9,10(*n*)-^{3}H]Octadec-9-en-1-ol	d_6-benzene	5.490	9,10	89–96	C	[119]
$\alpha'CH_2OCO(CH_2)_{14}CH_3$ $\beta CHOCO(CH_2)_{14}CH_3$ $\alpha CH_2OP(=O)(O^-)OCH_2CH_2\overset{+}{N}(CH_3)_3$ L-α-Phosphatidyl[N-*methyl*-^{3}H]choline dipalmitate	d_6-DMSO	3.08	*N*-methyl	100	A	[23]

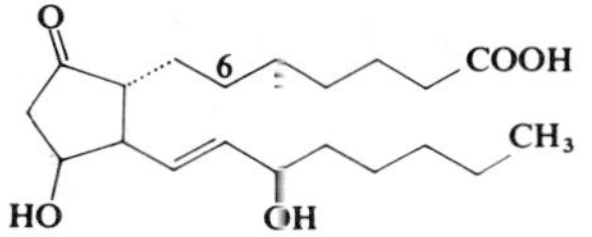 [5,6-^{3}H]Prostaglandin E_1	$CDCl_3$	1.28 1.32	5 6	100	C	[23]
$H_2N.\overset{4}{C}H_2\overset{3}{C}H_2\overset{2}{C}H_2\overset{1}{C}H_2NH_2$ [1,4(*n*)-^{3}H]Putrescine dihydrochloride	D_2O	3.03	1,4	90.6	C	[23]
		1.74	2,3	9.4	C	[23]
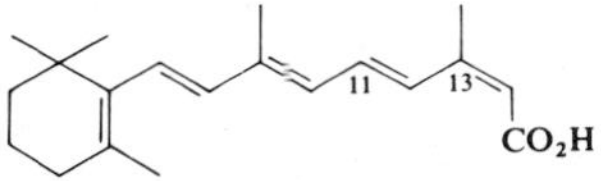 13-*cis*-[11-^{3}H]Retinoic acid	—	6.88	11	100	B	[120]
Sodium [^{3}H]acetate	D_2O	1.85	Methyl	100	J	[4]
[α,γ-^{3}H]Sorbic acid	d_6-DMSO	5.79 6.29	α-CH γ-CH	36 64	I	[109]
$CH_3(CH_2)_{15}\overset{\alpha}{C}H_2CO_2H$ [^{3}H]Stearic acid	$CDCl_3$	2.08	α-CH_2	100	H, I	[65]
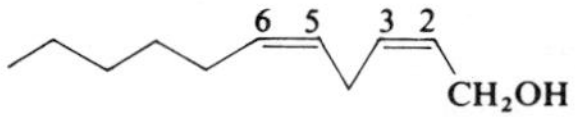 [2,3,5,6(*n*)-^{3}H]Undeca-2*Z*,5*Z*-dien-1-ol	CD_3OD	5.58 5.48 5.42 5.36	2 3 5 6	25 25 25 25	C	[23]

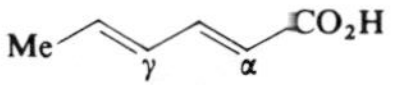

important for biochemical and biomedical tracer applications. These include, for example, [^{3}H]arachidonic acid and [^{3}H]eicosatrienoic acid, which are important intermediates in the biosynthesis of prostaglandins, and also [^{3}H]prostaglandin E_1, [^{3}H]stearic acid and L-α-phosphatidyldipalmitoyl[N-*methyl*-^{3}H]choline, a member of the important class of phospholipids involved in cell membrane studies. The tritium labelling data for these and other compounds are listed in Table 32. In a high proportion of the compounds, the specificity of labelling and the distribution patterns of tritium are as expected from the method of preparation: indeed, often the regiospecificity is 100 per cent.

CHAPTER 3

Applications of tritium nmr spectroscopy

In addition to the use of ^{3}H nmr spectroscopy for defining the patterns of labelling in tritium labelled compounds, many other applications of the nmr technique are developing in analysis, biochemistry, catalysis studies, environmental chemistry, radiation chemistry and in studies of reaction mechanisms. It is not necessary to read many publications on the applications of tritium labelled compounds in biological research to realize how essential it is to know the precise distribution of tritium in the compounds involved. This is especially true of all reactions involving enzymes, where hydrogen (tritium) atoms, not normally considered to be reactive, may be substituted or removed from the labelled molecules. Some examples follow.

The NIH-shift [121] would never have been discovered without clear evidence of the position of the tritium label in the starting material and product. For studies of the binding of [^{3}H]benzo[a]pyrene to biopolymers under the influence of microsomal enzymes, there is need to know the labelling pattern in this carcinogen [122, 123]. Equally, this knowledge is necessary for studying the metabolism of the carcinogen in cell cultures [28, 122]. Similarly, a detailed knowledge of the pattern of tritium labelling in estrogens is required for the correct interpretation of experiments on the binding of these steroid hormones to DNA in neoplastic cells [124]. Studies of the biosynthesis of collagen (from tritiated proline) and of penicillins using tritiated valine require an accurate knowledge of the positions and quantitative distribution of the tritium within the molecule, as do studies of the isomerization of L-valine to D-valine for incorporation into actinomycin D [125]. A knowledge of the stereospecificity of labelling in tritiated cholesterol is necessary for the study of 7α-hydroxylating enzymes in which the release of tritiated water by microsomal cholesterol 7α-hydroxylase gives a measure of the enzyme activity [126].

Other examples where knowledge of the precise labelling patterns in tritiated compounds is necessary include mechanistic studies on chemical transformations of tritiated substrates and the study of new labelling techniques, such as labelling by the microwave activation of tritium gas [127].

1. Analysis: structural analytical applications

By far the most important use of ^{3}H nmr spectroscopy is in the analysis of tritium labelled compounds to establish the distribution and stereochemistry of labelling, as indicated in Chapter 2. Synthetic procedures for tritium labelling not infrequently lead to an expected distribution of labelling, as illustrated in the Table 33. There are, however, numerous instances where this is not so. In either case, ^{3}H nmr is especially useful in quickly establishing the labelling pattern. This is simply because the technique is so direct and convenient, and shows both the regio- and stereospecificity of the tritium labelling, as well as the quantitative distribution of the tritium. Moreover, the nmr technique can often help in unravelling the nature of mixtures.

Table 33. Specificity of tritium labelling in some tritiated compounds

Compound class	Compound	Tritium in the position(s) stated (%)	Tritium in other positions (%) (position)
Amino acids	L-[3,4-^{3}H]Proline	95	4(5)
	L-[5-^{3}H]Tryptophan	99	
	L-[*ring*-2,6-^{3}H]Tyrosine	99	
Hydrocarbons	[6-^{3}H]Benzo[a]pyrene	94	4(1), 2(3)
		91	8(1), 1(3)
Nucleic acids	[8-^{3}H]Adenosine 3′,5′-cyclic monophosphate	99	
	[6-^{3}H]Thymidine	99	
	[*methyl*-^{3}H]Thymidine	99	
	[5-^{3}H]Uridine	99	
Steroids	[6,7-^{3}H]Estradiol	99	
	[1,2-^{3}H]Testosterone	99	
	[1,2-^{3}H]Progesterone	99	
	[2,4-^{3}H]Estriol	80	20(6,9)
	[7-^{3}H]Pregnenolone	80	20(4)
Miscellaneous	[*methyl*-^{3}H]Acetophenone	100	
	[α-^{3}H]Cinnamic acid	100	
	[2-^{3}H]Diethyl malonate	100	
	[*side-chain*-^{3}H]Phenylacetylene	100	

Labelled unsaturated aliphatic, aromatic and heterocyclic carboxylic acids can be prepared through the decarboxylation of the products of the Knoevenagel condensation between aldehydes and malonic acid [109]. Thus benzylidene-malonic acid, when heated in a tritiated water–dioxan mixture containing some pyridine, readily decarboxylated over the course of a few hours to give a good yield of the desired α-tritiated cinnamic acid, as demonstrated by its ^{3}H nmr spectrum (only one singlet at $\delta = 6.5$ p.p.m.). Hex-2-enoic acid and crotonic acid were similarly α-tritiated, as indicated:

$$RCHO + H_2C(CO_2H)_2 \xrightarrow[\text{base}]{\text{amine}} R.CH = C(CO_2H)_2$$

$$\downarrow -CO_2,\ THO$$

$$R.CH = CT.CO_2H$$

In the case of sorbic acid, however, prepared from the unsaturated aldehyde, crotonaldehyde, ^3H nmr spectroscopy showed that labelling had occurred at both the α- and γ-positions, i.e. the product was $CH_3.CH = CT.CH = CT.CO_2H$. This unexpected result is thought to arise because of a lactonization reaction, preceding the decarboxylation, giving an intermediate capable of base-catalysed exchange at the γ-carbon (see page 199). Similarly α,γ-labelling occurred in the product from cinnamylidenemalonic acid for the same reason. So even in tritium labelling procedures which do not involve a metal–hydrogen transfer catalyst, unexpected patterns of labelling may be observed when hitherto unexpected processes intervene.

Numerous tritium labelled compounds are of course prepared by reactions involving hydrogen transfer catalysts and in such cases some degree of non-specific labelling is usually observed. Thus in the catalytic addition of tritium across a double bond the product (as, for example, from oleic acid) may have tritium distributed along the hydrocarbon chain on either side of the originally sp^2-hybridized carbon atoms (see page 25). In other cases, the addition of tritium may be specific in the sense that it is confined to the carbon atoms originally joined by the double bond, but it may be distributed between these two positions very unevenly. For example, 4-amino-3-*p*-chlorophenyl-2,3-dehydrobutyric acid (2,3-dehydro-Baclofen) gives specifically 4-amino-3-*p*-chlorophenyl[2,3-^3H] butyric acid ([2,3-^3H]Baclofen) but there is twice as much tritium in the 3-position as in the 2-position [40].

An example where there is both uneven addition of tritium across a double bond and non-specific labelling (though confined to one extra site) is in the preparation of tritium labelled dihydroalprenolol, an important β-adrenergic antagonist. (−)Alprenolol (**XIX**) as the tartrate in water is reduced with tritium gas in the presence of 10 per cent. palladium-on-charcoal as catalyst [128]. The ^3H nmr spectrum of the product (**XX**) in deuterated DMSO showed that tritium was present in all three positions of the propyl side chain. The distribution ratio was 1:3.6:10 for the 1-methylene, 2-methylene and 3-methyl positions which

$OCH_2CH(OH)CH_2NHCH(CH_3)_2$, $CH_2CH = CH_2$ (on benzene ring) → $OCH_2CH(OH)CH_2NHCH(CH_3)_2$, $CHTCHTCH_2T$ (on benzene ring)

(**XIX**) (**XX**)

had chemical shifts of $\delta = 2.53$, 1.52 and 0.85, respectively [49, 128]. The small proportion of tritium found in the benzylic 1-methylene group was not really surprising, as it is known that such protons undergo hydrogen–tritium exchange under the catalytic conditions used [57]. Similar unsymmetrical addition of tritium to the double bond is observed in the preparation of [4,6,*propyl*-^{3}H]dihydroalprenolol (**XXI**) by simultaneous catalysed halogen–tritium replacement and double-bond reduction of 4,6-dibromoalprenolol [49]. In this case the ^{1}H-decoupled ^{3}H nmr spectrum (Figure 40) shows that 77 per cent. of the tritium is located in the propyl side chain in the ratio 1 : 4.7 : 10.6 for the 1-methylene, 2-methylene and 3-methyl groups, respectively, with the remaining 23 per cent. of the tritium located in the ring 4,6-positions. The reason for the unsymmetrical addition of tritium across a terminal double bond is now known to be concomitant catalysed hydrogen–tritium exchange at terminal vinylic positions, and this is discussed in detail on pages 191 to 193.

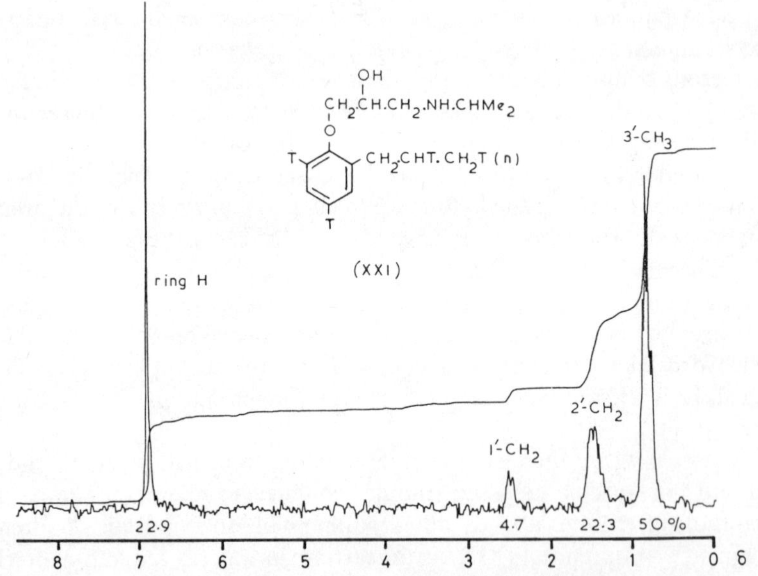

Figure 40. ^{3}H Nmr spectrum (^{1}H-decoupled) of [4,6,*propyl*-^{3}H]dihydroalprenolol in d_6-DMSO

An important consequence of the development of ^{3}H nmr spectroscopy as an analytical tool is that it has made the study of catalysed hydrogen isotope exchange reactions considerably easier than hitherto, as will have become apparent from preceding examples. Of course, mass spectrometry and radio-gas chromatography are frequently used as analytical tools for the examination of tritiated compounds, but neither technique can give direct information on the stereochemistry of the label, although some positional information may be derived. The ^{3}H nmr method, on the other hand, is more direct, powerful and generally applicable. Although nmr spectroscopic methods lack very high sensitivity, ^{3}H nmr has such considerable advantages to offer over the analysis of ^{3}H compounds by conventional stepwise chemical or biochemical degradation and radioactivity measurement that the often necessary use of millicuries rather than microcuries (or less) of tritiated compounds as tracers may frequently be warranted.

The many examples quoted here (together with those given in Chapter 2) demonstrate that ^{3}H nmr spectroscopy is a widely applicable tool which has enabled the detailed analysis of many generally tritiated compounds to be achieved for the first time, e.g. generally labelled polycyclic aromatic hydrocarbons [27]. The ^{3}H nmr techniques enormously facilitate the examination of products from Wilzbach tritiation [1,49,56] and from hydrogen isotope exchange processes involving tritiated water or acids as the isotope source [1,35,27,62,68]. Also facilitated is the very close examination of the regio- and stereospecificity of tritiation methods believed to be specific [1,49], such as borotritide reductions [1,49,78] and decarboxylation in the presence of THO or other tritiating media [1,49,109].

^{3}H Nmr spectroscopy is undoubtedly the method of choice for elucidating the distribution of label in a wide range of organic compounds, both simple and complex.

2. Biochemistry

The experimental study of biosynthesis has developed largely through the application of tracer compounds labelled with ^{14}C or ^{3}H. More recently, the introduction of Fourier transform nmr techniques has greatly increased the practicability of ^{13}C as an isotopic label, and many studies, particularly concerning polyketides, terpenes and porphyrins, have been reported [129]. It would be logical therefore to expect ^{3}H nmr spectroscopy to be used in biosynthetic studies and already several examples have appeared [105, 115]. A first study [115] was on the incorporation of [^{3}H]acetate into penicillic acid (**XXII**; see Figure 42), a metabolite of *Penicillium cyclopium.* This was chosen because the biosynthesis had been extensively investigated with carbon isotopes and a direct comparison between established procedures and the new approach would emerge. In fact, more information came directly from the use of tritium labelled precursors and ^{3}H nmr spectroscopy than from the isotopic carbon studies, partly because the use of isotopic hydrogen will distinguish between non-equivalent hydrogen atoms at sp^2 (vinylic) methylene groups (or indeed at prochiral sp^3 sites).

The ^{3}H nmr spectrum (Figure 41a) of the derived penicillic acid was assigned from line multiplicities and by reference to the corresponding ^{1}H nmr spectrum (Figure 41b). Tritium was found to be present at the 3-, 5- and 7-positions, consistent with the mode of biosynthesis outlined in Figure 42. The 7-position showed greatest incorporation of tritium (least exchange loss of label), with the 3- and 5-positions having progressively less tritium, in agreement with chain

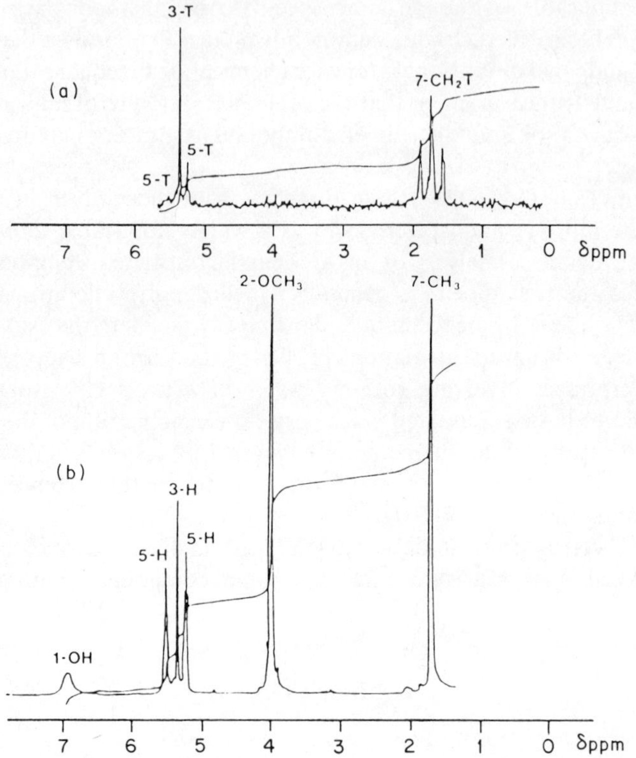

Figure 41. (a) Proton-coupled ^{3}H nmr spectrum of biosynthetic [^{3}H]penicillic acid in d_6-acetone and (b) the ^{1}H nmr spectrum. (Reproduced by permission of The Royal Society of Chemistry.)

initiation by an acetate methyl group rather than an activated chain-building methylene group derived from malonate via acetate. The 5-methylene group in penicillic acid had preferentially been labelled trans to the 7-methyl group as in (**XXII**). Subsequent incorporation studies using [$^{3}H_2$]malonate led to labelling of only the 3- and 5-positions of penicillic acid. In addition, [3,5-$^{3}H_2$]orsellinic acid (**XXIII**) was established as an advanced precursor by being incorporated into penicillic acid with the same distribution of label between the 5-trans- and 5-cis-positions as from the simpler precursors.

R_1 R_2 β α β α O

(**XXIV**) R_1 = OH, R_2 = H
(**XXV**) R_1R_2 = O

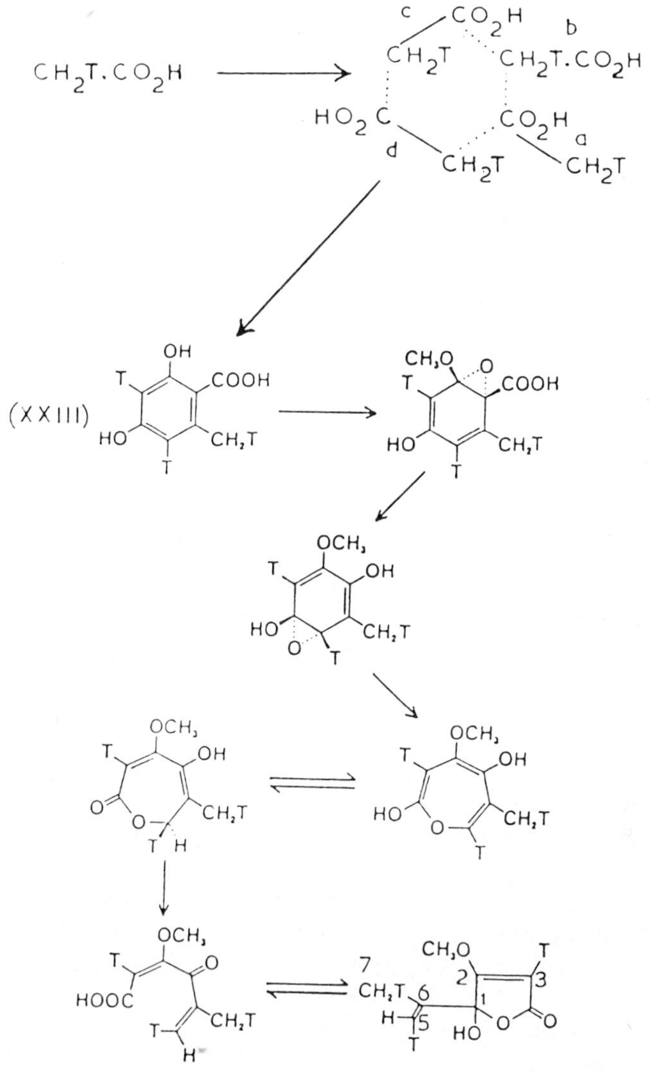

Figure 42. Possible biosynthetic scheme for penicillic acid (**XXII**) from acetate via orsellinic acid (**XXIII**), where a is the initiating acetate unit and b, c and d are chain-building units derived from malonate via acetate. (Reproduced by permission of The Royal Society of Chemistry.)

^{3}H Nmr spectroscopy has also been used to study the steric course of the biotransformation of testosterone (**XXIV**) to androsta-1,4-diene-3,17-dione (**XXVI**) in the bacterium *Cyclindrocarpon radicicola* [103]. The [1,2-^{3}H]testosterone employed consisted mainly of the two cis labelled species, namely the [1α, 2α-$^3H_2$] and [1β, 2β-$^3H_2$]compounds in the ratio of ca. 5:1. In this experiment,

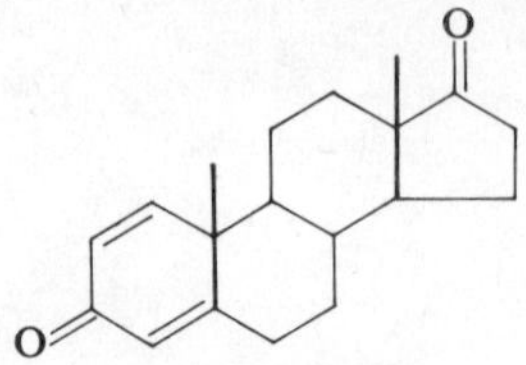

(XXVI)

45 mCi (0.25 mg) of **XXIV** was diluted with non-radioactive compound (25 mg) and, after incubation, androsta-1,4-diene-3,17-dione (**XXVI**) (23 mg, 14.5 mCi) was isolated for ^{3}H nmr examination. The ^{3}H nmr spectrum of the product showed that the signals at C-1 and C-2 had an intensity ratio of 1:5, thus directly demonstrating a 1α, 2β-trans elimination pathway, as previously reported for analogous dehydrogenations of 3-oxo-steroids by *Bacillus sphaericus* [130]. In a parallel experiment, the conversion of **XXIV** to **XXVI** was effected in a 45 per cent. yield by *Pseudomonas testosteroni* grown on a yeast medium and, again, complete 1α,2β-trans diaxial elimination was revealed by the ^{3}H nmr observations.

In contrast, an analogous dehydrogenation by 2,3-dichloro-5,6-dicyanobenzoquinone (DDQ) was less specific. The [1α,2α-$^3H_2$]3,17-dione (**XXV**) (29 mg, 26.5 mCi) which contained also one-sixth of [1β,2β-$^3H_2$] labelled compound was recovered as a by-product from the *P. testosteroni* experiment and then subjected to dehydrogenation with DDQ. This gave **XXVI** (12.6 mg, 4.2 mCi) which showed a major 2-^{3}H signal in the ^{3}H nmr spectrum (acquired with ^{1}H-decoupling), consistent with predominant 1α,2β-trans diaxial elimination. In addition, however, there were moderately intense 1-^{3}H and 2-^{3}H doublets arising from [1,2-$^3H_2$]androsta-1,4-diene-3,17-dione (**XXVI**) which could only have arisen from accompanying cis elimination processes. The DDQ dehydrogenation of predominantly [1β,2β-$^3H_2$]testosterone (**XXIV**) gave diene-dione (**XXVI**) for which the major signal came from 1-^{3}H as required by 1α,2β-trans elimination. Again, however, there were 1-^{3}H and 2-^{3}H doublets attributable to [1,2-$^3H_2$]diene-dione (**XXVI**), indicating that cis eliminations accompanied the predominant 1α,2β-trans diaxial elimination in the chemical dehydrogenation. These findings emphasized the complete stereo-specificity of the biochemical dehydrogenation process.

In another biochemical application [105] ^{3}H nmr spectroscopy has been used to elucidate the stereochemistry of the cyclization of 2,3-oxidosqualene (**XXVII**) to cycloartenol (**XXVIII**), a step in the biosynthesis (Figure 43) of sterols by photosynthetic organisms. The reaction involves the 1,3-cyclization of a steroid carbonium ion intermediate, through loss of a proton from a methyl group, to form the cyclopropane ring. The process could take place in a stereochemically defined way, involving either retention or inversion of configuration at the 19-methyl group (Figure 44). Through detailed study of line widths, long range couplings and induced shifts, it was possible to assign the cycloartenol ^{1}H resonances at $\delta = 0.16$ and 0.44 to the *exo*- and *endo*-cyclopropylmethylene

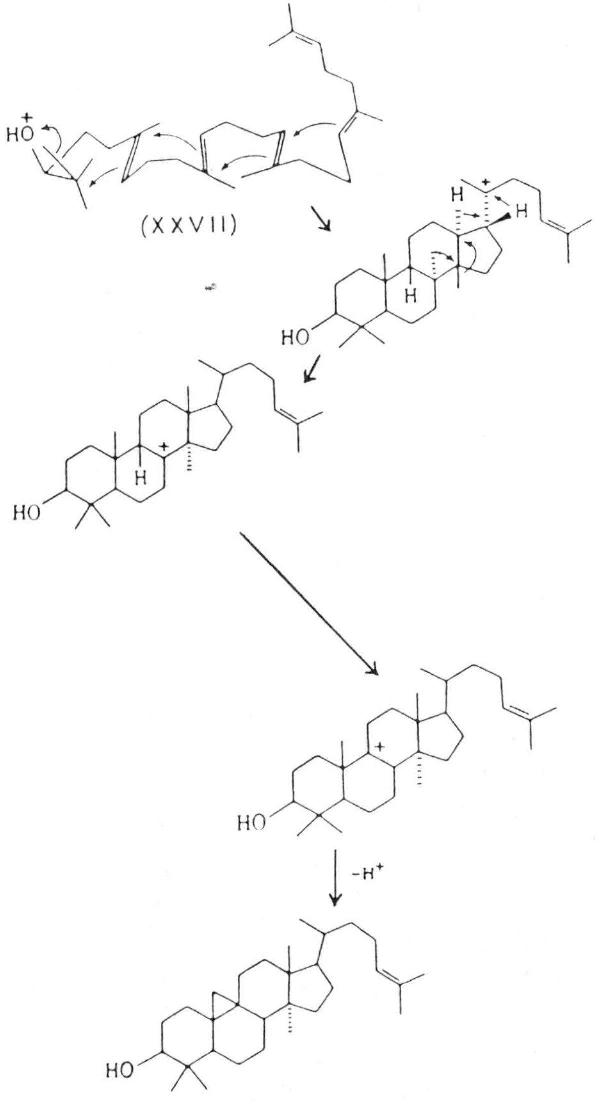

Figure 43. Steps in the biosynthesis of cycloartenol (**XXVIII**) from 2,3-oxidosqualene (**XXVII**) by photosynthetic organisms. (Reprinted with permission from *J. Am. Chem. Soc.*, **100,** p. 3236. Copyright 1978 American Chemical Society.)

protons, respectively. 2,3-Oxidosqualene with a chirally labelled (*R*)-CHDT methyl group was then synthesized from D-malic acid according to the scheme shown in Figure 45. In the intermediate product (**XXIX**), 30 per cent. of the molecules bore tritium in the methyl group and all those also carried one

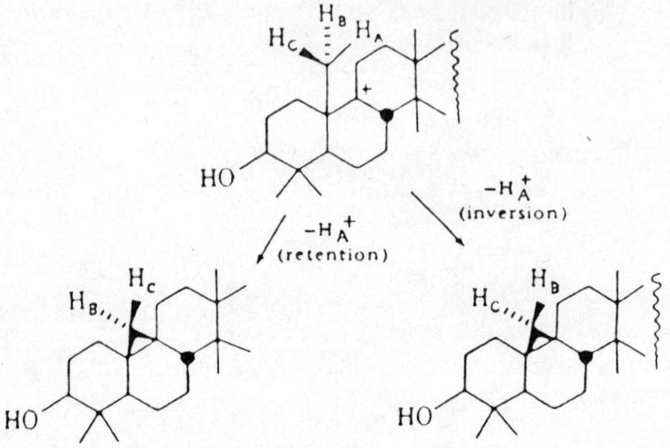

Figure 44. Possible ring-closure mechanisms in the formation of cycloartenol. (Reprinted with permission from *J. Am. Chem. Soc.* **100**, p. 3236. Copyright 1978 American Chemical Society.)

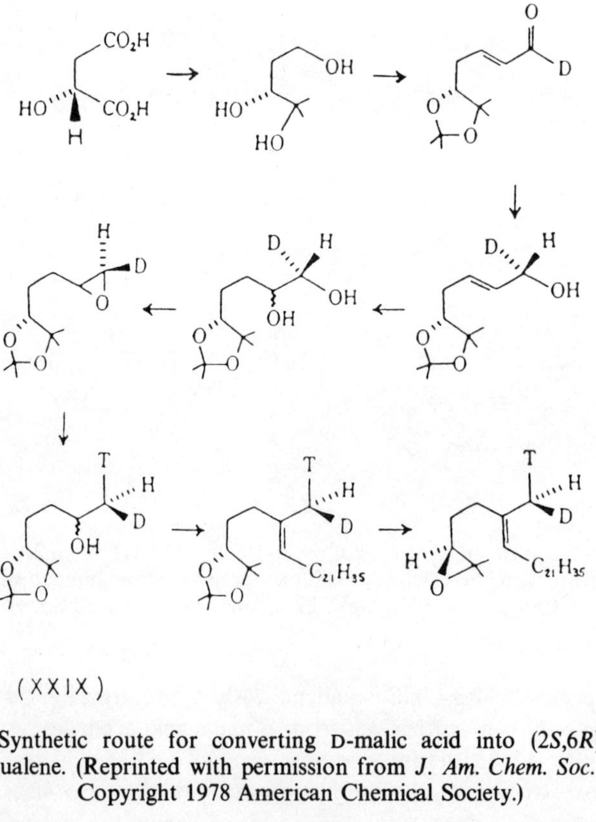

(XXIX)

Figure 45. Synthetic route for converting D-malic acid into (2*S*,6*R*)-2,3-oxido[6-^{3}H,^{2}H,^{1}H]squalene. (Reprinted with permission from *J. Am. Chem. Soc.* **100**, p. 3237. Copyright 1978 American Chemical Society.)

deuterium and one hydrogen atom as shown by the ^3H nmr spectra with and without ^1H-decoupling (Figure 46). The conversion of the chirally methyl labelled oxidosqualene into tritium labelled cycloartenol was accomplished in a ca. 22 per cent. yield by incubation with a cell-free microsomal fraction from *Ochromonas malhamensis*. The ^3H nmr spectrum of the product showed a major singlet at $\delta = 0.168$, both with and without ^1H-decoupling, indicating the presence of an *exo*-cyclopropyl triton in molecules having an *endo*-cyclopropyl deuteron (no resolved splitting). There was a lesser singlet at $\delta = 0.456$ assignable to an *endo*-cyclopropyl triton in molecules having an *exo*-cyclopropyl proton, as shown by the doublet splitting in the absence of ^1H noise decoupling. It is thus clear that the conversion has taken place with retention of configuration, mainly through proton loss but to some extent through deuteron loss.

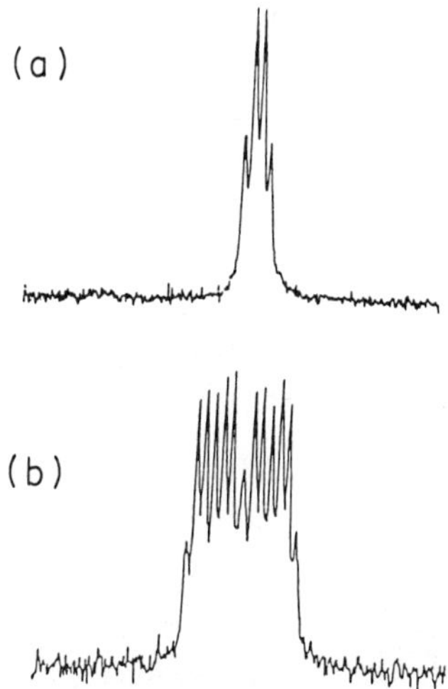

Figure 46. ^3H Nmr spectra of the intermediate (**XXIX**) (a) with and (b) without ^1H-decoupling. Compound (**XXIX**), being a diastereotopic pair, gives two triton signals, each of which is split into three equal lines by the coupling to the deuteron; overlap results in the observed 1:2:2:1 multiplet, (a). In (b) there is, additionally, *geminal*-coupling from a proton and also lesser *vicinal*-coupling from a proton; overlap results in the observed line pattern, (b). (Reprinted with permission from *J. Am. Chem. Soc.* **100,** p. 3236. Copyright 1978 American Chemical Society.)

Compounds containing chiral methyl groups are finding increasing application in studies of biosynthesis [131, 132], and this necessarily implicates the use of tritium. For ^3H nmr spectroscopy to be useful it is clear that three conditions have to be fulfilled:

(a) The chiral methyl compound used as substrate needs to be synthesized in such a way that it contains an adequate isotopic abundance of tritium and in most cases ca. 100 mCi to several curies will be required.

(b) The biosynthetic yields should be sufficiently good so as to provide enough tritiated product for ^3H nmr examination (see page 16).

(c) The signals of the relevant hydrogens must be adequately resolved in the spectrum and unambigously assigned.

Fortunately, since ^3H chemical shifts are essentially identical with ^1H or ^2H chemical shifts, the assignment is usually achieved by reference to the ^1H nmr spectrum of unlabelled material. In complex cases, the assignment problem may be solved by specific ^1H-decoupling experiments, through use of Overhauser effects or by means of specifically labelled compounds—either deuterated or tritiated. The requirements (a) and (b) are of course interdependent. Where it is possible to grow cells to maximum turnover and then transfer these to a minimal medium containing the required labelled substrate, surprisingly good incorporation (several per cent.) may be achieved [115], minimizing the total radioactivity needed.

The foregoing criteria (a to c) were all satisfied in a ^3H nmr study of the stereochemistry of the dehydrogenation of isobutyryl coenzyme A (**XXX**) to methylacrylyl coenzyme A (**XXXI**) [37]. This reaction, which is a step in the

$$(CH_3)_2CH\text{-}COS\overline{\text{CoA}} \longrightarrow CH_2{=}C(CH_3)\text{-}COS\overline{\text{CoA}}$$

(XXX) **(XXXI)**

catabolism of valine, is catalysed by a flavin-dependent enzyme similar to the general acyl-$\overline{\text{CoA}}$ dehydrogenase. Rather than the purified dehydrogenase, the bacterium *Pseudomonas putida* was used, which is capable of growing with isobutyric acid as the sole carbon source, and which accumulates β-hydroxyisobutyric acid in substantial yield. The required chiral methyl substrate was synthesized by reduction of stereospecifically labelled benzyl methacrylate (**XXXII**), followed by alkaline hydrolysis to **XXXIII** and **XXXIV**. Thus the product contains an (*S*)-methyl group at the 2-*pro*-(*S*)-position and an equal amount of an (*R*)-methyl at the 2-*pro*-(*R*)-position. The latter component can be ignored since the 2-*pro*-(*R*)-methyl group is not oxidized by the enzyme, but it will be carried through the metabolism and appear in the nmr spectra of the product as having a tritiated and deuterated methyl group. The substrate, comprising **XXXIII** and **XXXIV** (ca. 2 Ci), was then incubated with *Pseudomonas putida* and the resultant β-hydroxyisobutyric acid isolated in ca. 1 per cent. overall yield as the methyl ester benzoate whose ^1H and ^3H nmr spectra (Figure 47) were taken in the presence of adequate $Eu(fod)_3$ to resolve the methyl signals of interest. A major tritium signal at $\delta = 5.80$ (3-*pro*-(*S*)-H) and a lesser tritium signal at $\delta = 5.96$ (3-*pro*-(*R*)-H) resulted from metabolism of the *pro*-(*S*)-methyl of **XXXIII**, whereas the signal at $\delta = 1.74$ was derived from the unoxidized 2-*pro*-(*R*)-methyl of **XXXIV**. The relative signal intensities at $\delta = 5.80$ and 5.96 were consistent

(XXXIII) ← (a) T_2 (b) OH^- — (XXXII) — (a) T_2 (b) OH^- → (XXXIV)

(c), (d), (e)

(c) dehydrogenation } microbial
(d) hydration
(e) benzoylation, esterification

with expectations for an anti-elimination of hydrogen from **XXXIII**, assuming $k_H > k_D > k_T$.

As a preliminary to other biochemical studies, use of ^3H nmr spectroscopy has also been made in elucidating the stereochemistry of the hydrogenation by Wilkinson's catalyst of deuterium labelled *N*-acetylisodehydrovaline (**XXXV**)

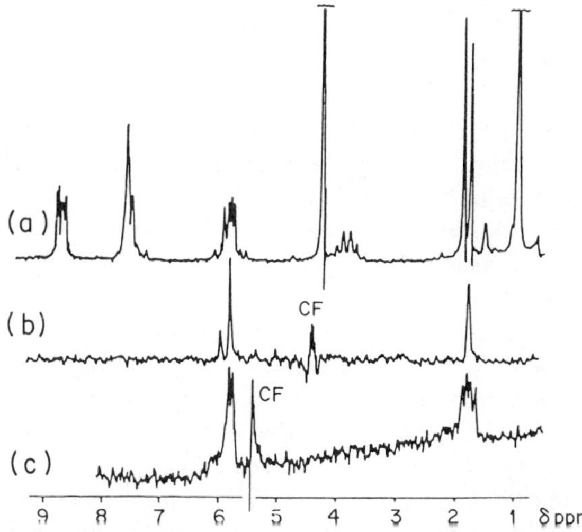

Figure 47. ^3H Nmr spectra of methyl β-benzoyloxyisobutyrate in $CHCl_3$: (a) ^1H-decoupled, (b) ^1H-decoupled + $Eu(fod)_3$ and (c) ^1H-coupled + $Eu(fod)_3$: CF = centre frequency of spectrometer (pulse point). (a), (b) reprinted with permission from *J. Am. Chem. Soc.*, **102**, p. 6377 (1980). Copyright 1980 American Chemical Society. (c) reproduced by permission of D. J. Aberhart.)

with tritium–hydrogen (TH, H_2) mixtures [39, 133]. The two diastereoisomeric pairs of products from the addition of HT were formed in unequal amounts (19:1). This showed that there was a strong preference for 3-*re*,4-*si* addition to the (*S*)-component of **XXXV** and 3-*si*,4-*re* addition to the (*R*)-component, as indicated below.

(*S*) CO_2H / AcNH—H / 3 / H—4 Me / D (**XXXV**) (*R*) CO_2H / H—NHAc / 3 / Me 4 H / D

3-*re*, 4-*si* addition — HT/catalyst — 3-*si*, 4-*re* addition

CO_2H / AcNH—H / Me—H / CHDT (*R*) CO_2H / H—NHAc / H—Me / CHDT (*S*)

Preponderent (2*SR*, 3*SR*, 4*RS*)-[4-$^3H^2H^1H$]-*N*-acetylvaline

Chiral methyl valine was used in the elegant work [132] on the stereochemistry of the incorporation of the methyl groups of valine into the methylene groups of cephalosporin C (**XXXVI**), according to the scheme shown in Figure 48. In this scheme the 4-*pro*-(S)-methyl group of valine becomes the *exo*-cyclic methylene (C-3′) and the 4-*pro*-(*R*)-methyl group becomes the *endo*-cyclic methylene group C-2 (indicated by * and ‡, respectively, in Figure 48). In this study an equimolar mixture of 108 mCi of (2*S*, 3*S*, 4*R*)- and (2*S*, 3*R*, 4*S*)-[4-^3H,^2H,^1H]valine (**XXXVII**) was incorporated into cephalosporin C by the use of a suspension of washed cells of *Cephalosporium acremonium*. The ^1H and ^3H (^1H-decoupled) nmr spectra of the derived cephalosporin C (250 μCi) are shown in Figure 49a and b, respectively.

In the ^3H nmr spectrum (see Figure 49b), the region corresponding to the *exo*-cyclic methylene group C-3′ of cephalosporin C showed a major singlet at $\delta = 4.4$ p.p.m. and a minor singlet at $\delta = 4.6$ p.p.m. An undecoupled spectrum was obtained, the signal-to-noise ratio of which only permitted observation of the signal at $\delta = 4.4$ p.p.m., which, however, remained as an apparent singlet. This signal, therefore, must have arisen from a CDT species. This result is as expected for a stereospecific conversion in which a normal deuterium isotope effect operated, leading to a predominance of the CDT species produced by cleavage of a C–H bond, over the corresponding CHT species produced by cleavage of a C–D bond. In the region corresponding to the *endo*-cyclic methylene group, singlets at $\delta = 3.4$ and 3.1 p.p.m. were observed of approximately 2:1 relative intensity.

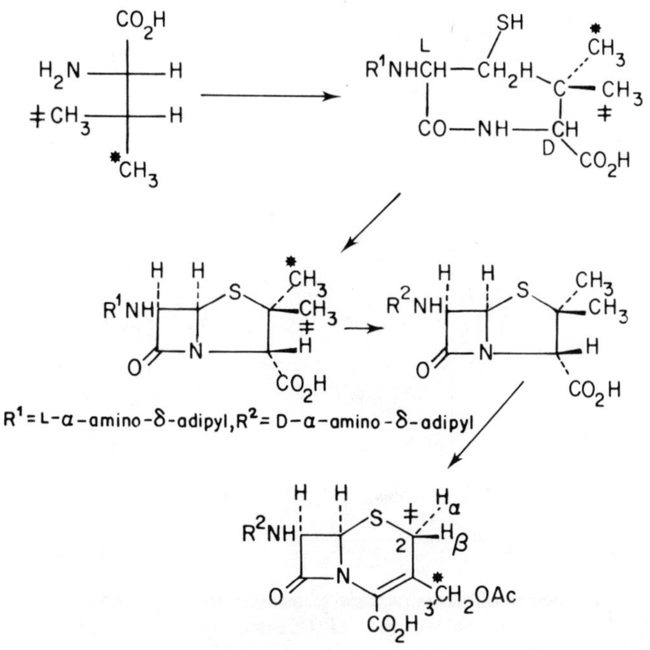

CO$_2$H, H$_2$N, H, ‡CH$_3$, H, *CH$_3$ → R^{1}NHCH—CH$_2$H—C(*CH$_3$)(CH$_3$‡), L, SH, CO—NH—CH, D, CO$_2$H

R^1 = L-α-amino-δ-adipyl, R^2 = D-α-amino-δ-adipyl

(XXXVI)

Figure 48. Conversion of valine into cephalosporin C by *Cephalosporium acremonium*. (Reproduced by permission of The Royal Society of Chemistry.)

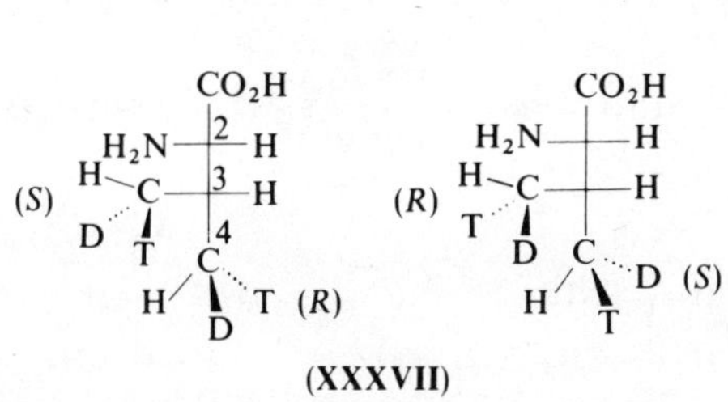

(XXXVII)

In a corresponding experiment, 1.45 Ci of an equimolar mixture of (2*S*,3*S*,4*S*)-, (2*R*,3*S*,4*S*)-, (2*R*,3*R*,4*R*)- and (2*S*,3*R*,4*R*)-[4-^{3}H,^{2}H,^{1}H]valine (**XXXVIII**) was used to obtain tritiated cephalosporin C (1.2 mCi). The ^{1}H-coupled ^{3}H nmr spectrum of the cephalosporin C produced is shown in Figure 49c, the broad-band proton-decoupled spectrum being essentially the same as in Figure 49b. The spectrum (Figure 49c) shows a labelling of the *exo*-cyclic methylene group of cephalosporin C complementary to that found in the first experiment, consisting of an apparent singlet, attributable to a CDT group at $\delta = 4.6$ p.p.m. (In this experiment, no related minor signal at $\delta = 4.4$ p.p.m. could be seen.) It is estimated that if any of the CHT species had been present, the corresponding doublet would have been discernible if the intensities of the components of the

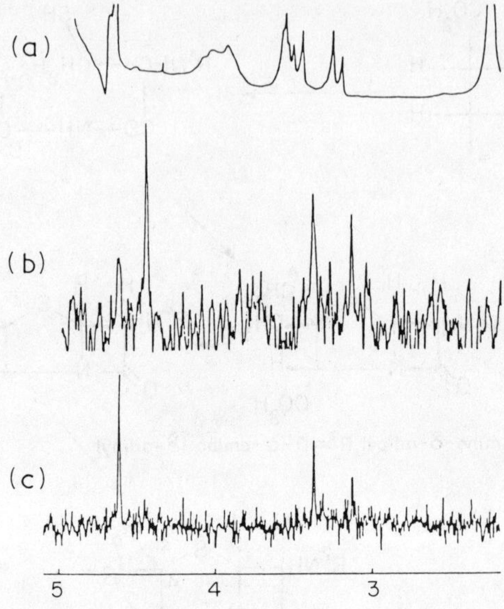

Figure 49. Nmr spectra of cephalosporin C, as sodium salt in D_2O, derived from chiral methyl valine (**XXXVII**): (a) ^1H, (b) ^3H (^1H-decoupled) and (c) ^3H (^1H-coupled). (Reproduced by permission of the Royal Society of Chemistry.)

CO_2H
H_2N—$\!$—H
H_3C—$\!$—H
CHDT (*S*)

CO_2H
H—$\!$—NH_2
H_3C—$\!$—H
CHDT (*S*)

(XXXVIII)

CO_2H
H—$\!$—NH_2
H—$\!$—CH_3
CHDT (*R*)

CO_2H
H_2N—$\!$—H
H—$\!$—CH_3
CHDT (*R*)

doublet had exceeded 20 per cent. of the intensity of the signal due to the CDT species. The results with respect to the *exo*-cyclic C-3′ methylene group are therefore entirely in accord with the expectation of a stereospecific oxidation with a stereospecificity exceeding 70 per cent. Since the signals due to this methylene group have not been assigned, it is not possible at present to state whether this result corresponds to retention or inversion of configuration.

This study was especially interesting in demonstrating how ^3H nmr spectroscopy, together with the use of configurationally labelled substrates, will permit simultaneous observation of stereochemical events at two different centres in a biosynthetic transformation.

3. Catalysis

In the development of modern theories of catalysis, hydrogen isotope exchange and hydrogenation reactions have played an important part [134]. In view of the ability of ^{3}H nmr spectroscopy to delineate the pattern of labelling in virtually any organic compound at the 0.1 to 10 mCi level of radioactivity, the technique is ideal for studying various aspects of catalysis as already adequately evinced by the many publications.

The capability of ^{3}H nmr spectroscopy for determining hydrogen isotope orientation in a tritium labelled compound after, say, a metal catalysed exchange reaction is fundamentally important for delineating the mechanism of the process. From such studies information on the nature of the interaction between compound and catalyst, either homogeneous or heterogeneous, can be inferred. These studies may also throw light on the relationship between homogeneous and heterogeneous catalysis. Recent work has shown that the manner of bond formation involving adsorbed molecules at a metal surface and the chemistry of inorganic coordination complexes are closely related [135].

Prior to the development of ^{3}H nmr spectroscopy there were several methods for measuring the distribution of the hydrogen isotope within a labelled compound, in addition to the traditional chemical degradation and counting (which is time-consuming and can lead to ambiguous results). ^{1}H Nmr spectroscopy coupled with mass spectrometry of deuterated compounds was one of these but since the position of deuterium in a molecule is indicated by the decrease of a signal in the ^{1}H nmr spectrum high levels of deuterium incorporation are necessary. This may be no great problem but difficulties certainly arise when the peaks in the ^{1}H nmr spectrum are poorly resolved. In certain circumstances ^{2}H nmr spectroscopy [136] may be employed but its intrinsic lower sensitivity and the poorer spectral dispersion compared with ^{1}H and ^{3}H nmr spectroscopy, together with the rather broader signals and the virtual absence of all but geminal spin-coupling information, may be strongly disadvantageous. For examination of generally labelled compounds for which the hydrogen nmr signals often occur in close groups, use of deuterium and ^{2}H nmr can be quite unhelpful compared with the use of tritium and ^{3}H nmr with ^{1}H-decoupling. The latter technique normally gives good resolution of the lines which can then be integrated satisfactorily.

Although electron spin resonance spectroscopy has been used [137, 138] in determining isotope orientation in deuterium exchange the types of compounds that can be analysed are somewhat limited. Infrared techniques for general deuterium orientation measurements have been used [139] in the past but this procedure is now known to be unreliable because of the problems associated with the presence of multiply deuterated species and the difficulty of assigning one band in the spectrum to represent quantitatively the isotope at a particular position. It is not surprising, therefore, that some of the early applications of ^{3}H nmr spectroscopy to studies in catalysis were concerned with reinterpreting existing data. An example concerns the powerful Lewis acid ethylaluminium dichloride [140 to 142] which in terms of rapidity of reaction, selectivity for aromatic ring positions and freedom from steric hindrance is one of the most attractive of catalysts for labelling a wide range of compounds with hydrogen isotopes derived from labelled water. Although aromatics are labelled preferentially, nevertheless aliphatic compounds such as alkanes and alkenes can also be labelled provided they contain a tertiary hydrogen atom [143]. The extent and pattern of deuterium incorporation (which was high) was found by a combination of ^1H nmr spectroscopy and mass spectrometry, whilst the tritium incorporation (necessarily low) was ascertained by radio-gas chromatography. The ^3H nmr results [68] for three alkanes (Table 34) confirm that the ethylaluminium dichloride catalyst tritiates the compounds efficiently but show that the label is not confined to the tertiary positions. The ^3H nmr spectrum (Figure 50) of the tritiated methylcyclohexane shows single methyl and CH-group chemical shifts for the equatorially methyl-substituted molecule and at least seven distinct shifts from the ten non-equivalent CH_2-group tritons.

Table 34. Position and extent of labelling in some alkanes using ethylaluminium dichloride and then THO. (Reprinted with permission from *Tetrahedron Lett.*, p. 4349 (1977). Copyright 1977, Pergamon Press)

Compound*	^3H Chemical shifts (p.p.m.)	Position labelled	Relative incorporation (%)	Relative incorporation per site
2,3-Dimethylbutane[a]	1.27	CH	17	8.5
	0.84	CH_3	83	6.9
3-Methylpentane[b]	1.28	CH	21	21
	1.10	CH_2	29	7.2
	0.84	CH_3	50	5.6
Methylcyclohexane[c]	1.63	CH	28	28
	ca. 1.26	CH_2	32	3.2
	0.84	CH_3	40	13.3

* Duration of exchange at 85 °C: (a) 4 days, (b) 1 day, (c) 19 days.

For studying the selectivity of catalysts, such as metals of the Group VIII series which catalyse the isotopic hydrogen exchange reaction:

$$RH + H^*OH \overset{M}{\rightleftharpoons} RH^* + H_2O \qquad (11)$$

^3H nmr spectroscopy proves ideal [93]. Results have shown that pre-reduced PtO_2 in this procedure labels toluene and butylbenzene both in the ring and in the side chain, whereas Raney nickel effects incorporation exclusively in the methyl group of toluene and the α-CH_2 group of butylbenzene. When the behaviour of pre-reduced PtO_2 is compared with that of Lewis acid-type catalysts such as ethylaluminium dichloride and boron tribromide, the differences are most marked in sterically hindered compounds such as 1,3,5-trimethylbenzene. The

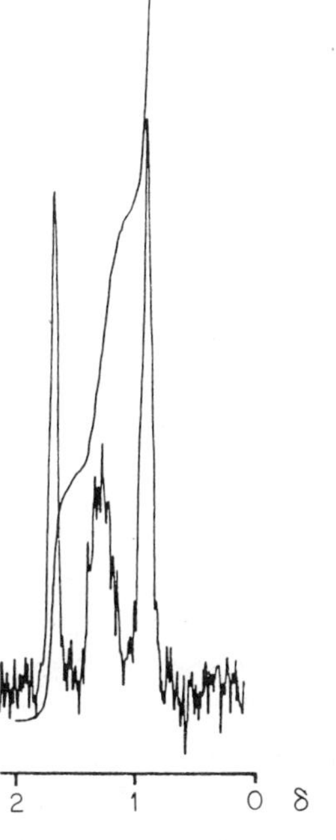

Figure 50. ^3H Nmr spectrum (^1H-decoupled) of [^3H]methylcyclohexane (neat). (Reprinted with permission from *Tetrahedron Lett.*, p. 4349 (1977). Copyright 1977, Pergamon Press.)

platinum catalyst leads to exclusive methyl group labelling and the other two catalysts to exclusive ring labelling. Interestingly the homogeneous catalyst, rhodium trichloride, gives similar results to the Lewis acid catalysts. The homogeneous tris-(triphenylphosphine)ruthenium(II) catalyst effected nearly specific exchange in primary alkanols at the α-methylene group (Table 35). For specific exchange it was important to keep the reaction time short because prolonged heating resulted in the β-methylene also becoming labelled. This

Table 35. Tritium labelling in alkanols with $(Ph_3P)_3RuCl_2$ catalyst. (Reproduced by permission of The Royal Society of Chemistry.)

Alkanol	Position	^3H Chemical shift δ (p.p.m.)	Relative incorporation (%)*
Ethanol	CH_2	3.1	100
Heptanol	α-CH_2	3.6	100[a]; 86[b]
	β-CH_2	1.5	0[a]; 14[b]
Benzyl alcohol	CH_2	4.0	100
3-Phenylpropanol	α-CH_2	3.1	96
	β-CH_2	1.4	4

* Results (a) after 0.5 h; (b) after 2 h.

finding, together with the failure of tertiary alcohols such as adamantanol to exchange, strongly implies an oxidation–reduction mechanism.

These one-step catalytic procedures when they do lead to exclusive labelling provide an attractive alternative to some of the more involved synthetic routes.

Evidence in support of a relationship between homogeneous and heterogeneous metal catalysis has come in part from a comparison of the orientation of the isotope in mono-substituted benzenes under these different conditions [144]. For such a study to be effective it is necessary to be able to observe orientation of exchange in the heterogeneous system at low levels of isotope incorporation, i.e. before the onset of scrambling. In the case of deuteration studies the levels of deuterium incorporation necessary for satisfactory analysis are precisely those under which scrambling can take place, leading to enhanced ortho-incorporation of isotope in monosubstituted benzenes. Some of the early conclusions [139] are therefore not valid.

^{3}H Nmr spectroscopic studies [135, 145] of the orientation of tritium in halogenated benzenes using heterogeneous platinum exchange show (Table 36) that as the size of the halogen increases the ortho/meta tritium ratio decreases. Electronic effects seem to be unimportant and the predominance of steric effects is consistent with the participation of the heterogeneous π-dissociative exchange mechanism (Figure 51) in preference to the analogous heterogeneous π-associative process. These conclusions are also consistent with the deactivation effects observed in the homogeneous deuterium exchange of monohalogenated benzenes with soluble $PtCl_4^{2-}$ as catalyst, where no competing scrambling occurs [135]. Hence the homogeneous π-dissociative mechanism (Figure 52) has been proposed for the exchange mechanism.

Data for alkylbenzenes (Table 36) show the same trend, i.e. the ortho/para ratio decreases as the alkyl size increases, and provide important information about the

Table 36. Orientation of tritium in halogenated and alkyl benzenes using heterogeneous platinum exchange.* (Reproduced by permission of The Royal Society of Chemistry.)

	Percentage of tritium in compound per site†			
Compound	Ortho	Meta	Para	Alkyl
Fluorobenzene	8.3	29	25	
Chlorobenzene	4.0	30	32	
Bromobenzene	< 2	33	33	
Toluene	9.8	15	18	CH_3(11)
Isopropylbenzene	5.5	13	13	CH_3(7), CH(11)
Cyclohexylbenzene	4.1	19	21	CH_2(1), CH(25)
Triphenylmethane	2.3	9.5	9.7	CH(< 1)

* Reaction conditions: unsupported platinum (0.1 g) obtained by sodium borohydride reduction of PtO_2, organic compound (0.1 ml), THO (0.1 ml, 5 Ci ml^{-1}), heated to 100–130 °C for an appropriate time in sealed ampoule (5 ml).

† Total activities of products typically 30–100 Ci mol^{-1}.

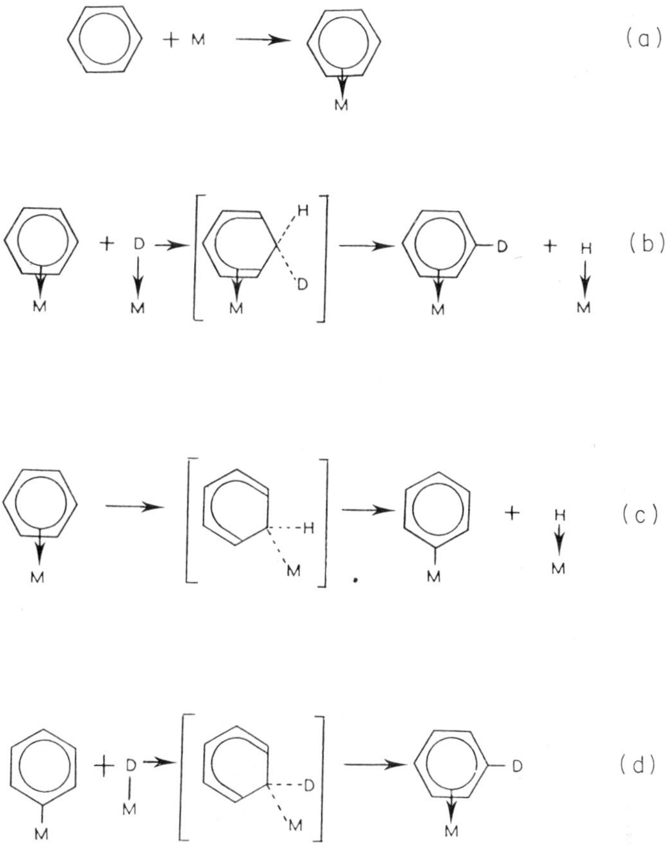

(a)

(b)

(c)

(d)

Figure 51. Heterogeneous associative and dissociative π-complex substitution mechanisms: associative mechanism, steps (a) and (b); dissociative mechanism, steps (a), (c) and (d). (Reprinted from *Catalysis Rev.*, **5**, p. 237 (1971) by courtesy of Marcel Dekker, Inc.)

magnitude of the steric effect for exchange in the side chain. Thus the methylene groups in cyclohexylbenzene show little isotope incorporation, indicating that the molecule is preferentially adsorbed through the benzene ring, and must be strongly tilted during this π-complex interaction such that aromatic and allylic positions can exchange preferentially. In triphenylmethane, even the benzylic position is strongly hindered in adsorption by the three aromatic rings and tritiation by a 'π-allylic' type of mechanism is precluded.

The data in Table 36 also show that the rate of isotope incorporation is approximately the same in the meta as in the para position for all seven monosubstituted benzenes exchanged over heterogeneous platinum. These results also support the view that the participation of a π-dissociative process in the heterogeneous exchange mechanism is predominant.

In connection with the measurement of the tritium distribution in generally labelled alkyl benzenes it is pertinent to mention a potential source of

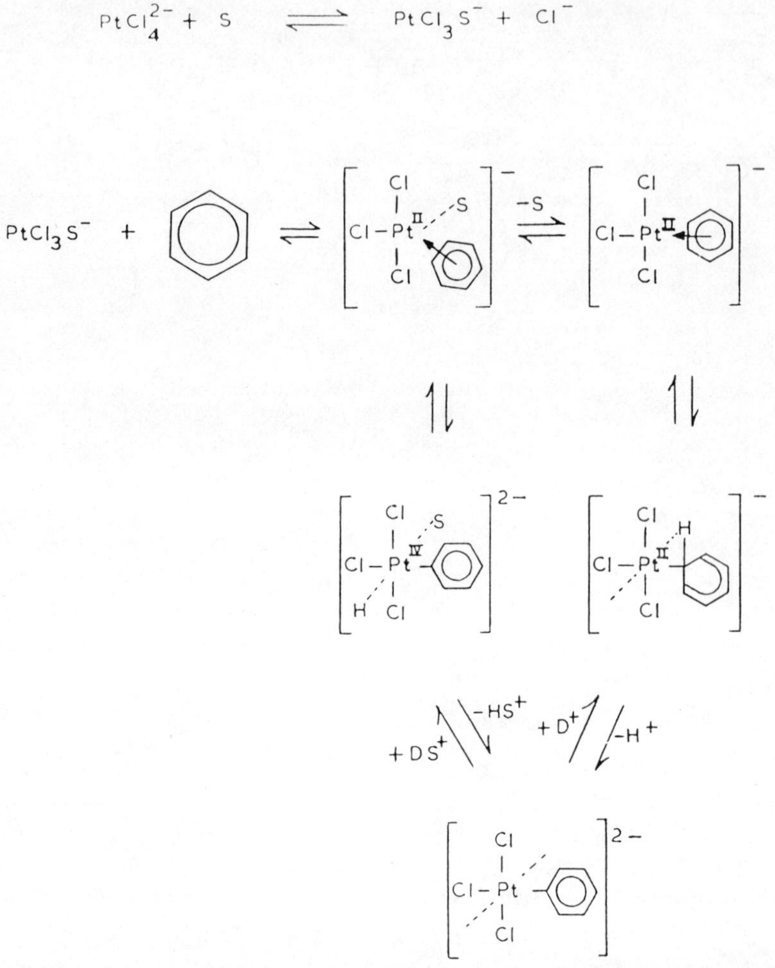

$$PtCl_4^{2-} + S \rightleftharpoons PtCl_3S^- + Cl^-$$

Figure 52. Mechanism for benzene hydrogen exchange with homogeneous platinum catalysis. Reprinted from *Chemistry in Australia*, **47**, 218 (1980) by permission of the Royal Australian Chemical Institute.

misinterpretation [34, 95, 97]. The need to maintain as high a sensitivity as possible in ^3H nmr analysis may lead to the use of pure compounds or concentrated solutions in the spectrometer sample. Identification of the labelled sites would then have to be made by reference to ^1H chemical shifts determined similarly. It is not safe to assume, for example, that close chemical shifts will have the same relative order for neat liquids as for dilute solutions (see pages 4, 5). These matters are very well recognized by nmr spectroscopists, but not necessarily by others interested in making use of the technique. Table 37 lists the aromatic ^3H chemical shifts for neat toluene and for a solution, as measured on a Bruker CXP-300 spectrometer operating at 320 MHz. Shifts were measured

Table 37. ^{3}H Chemical shifts in toluene (p.p.m.). (Reproduced by permission of John Wiley & Sons Ltd.)

Toluene concentration	δ_{ortho}	δ_{para}	δ_{meta}	δ_{methyl}
100%	7.013	7.041	7.123	2.117
10% in CCl_4	7.095	7.082	7.176	2.296
10% in DMSO	7.214	7.190	7.291	2.282
10% in cyclohexane	7.069	7.050	7.142	2.236

relative to internal tetramethylsilane by the ghost referencing procedure. Reversal of the ortho and para shifts occurs as the solvent composition is altered.

A number of Lewis acids, in addition to ethylaluminium dichloride, catalyse the hydrogen isotope exchange reaction. These include $SbCl_5$, $NbCl_5$, $AlCl_3$ and BBr_3. When the last catalyst is used to tritiate some substituted aromatics, products of high specific activity are obtained [35]. Furthermore, the tritium is directed to those positions in the compound which are favoured in electrophilic substitution. This is in marked contrast to results with $EtAlCl_2$ where labelling within the aromatic nucleus appears to be random (Table 38). These differences may reflect a difference in the strength of the charge transfer interaction between the ring and catalyst.

Table 38. Tritiation of substituted aromatics.* (Reprinted with permission from *Tetrahedron, Lett.*, 4171 (1978). Copyright 1978, Pergamon Press.)

Compound	Catalyst	Activity ($Ci\ mol^{-1}$)	Tritium incorporation (position and relative percentage per site)			
			Ortho	Meta	Para	Alkyl
Toluene	BBr_3	60	30	< 10	40	< 3
Isopropylbenzene	BBr_3		18	5	19	CH(16), CH_3(6)
Bromobenzene	BBr_3	18	26	< 3	47	
Naphthalene	BBr_3	40	α,19	β,6		
Bromobenzene	$EtAlCl_2$	25	19	18	25	
Bromobenzene†	Pt	19	< 2	33	33	

* Reaction conditions: 0.1 ml organic, 0.05 g catalyst and 0.5 ml HTO ($5\ Ci\ ml^{-1}$) in a sealed tube at 70°C for 12 h.

† 170 °C for 45 h.

The pattern of tritium distribution achieved by Lewis acid catalysts also contrasts markedly with that typical of heterogeneous platinum exchange. Indeed, the extreme deactivation of the ortho positions in the latter catalysis would make the following specific exchange possible [35]:

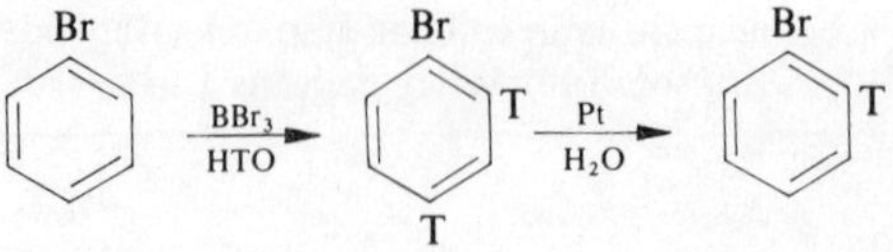

Br Br Br

$\xrightarrow[HTO]{BBr_3}$ T $\xrightarrow[H_2O]{Pt}$ T

T

Studies of metal-catalysed hydrogen isotope exchange in heterocyclic compounds have contributed greatly to our present understanding of mechanisms on metal surfaces [135]. In two separate studies [108,146] ^{3}H nmr spectroscopy has been used to analyse the tritium distribution in a series of nitrogen-containing heterocyclic compounds subjected to exchange (with tritiated water) catalysed by pre-reduced PtO_2. In the first of these studies [108], in which exchange was allowed to proceed to equilibrium, there was severe steric hindrance to exchange in some of the compounds, e.g. at the 1- and 10-positions in phenanthridine (**XXXIX**) which contained less than 2 per cent. of the total incorporated tritium.

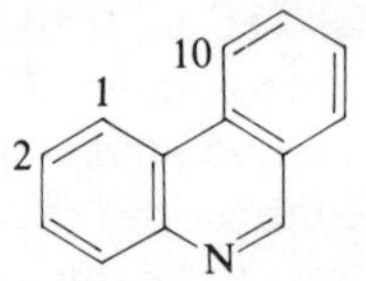

10
1
2
N

(XXXIX)

For this compound the ^{1}H nmr spectrum had not been fully analysed, so assignment of the ^{3}H chemical shifts was made by comparison with those for quinoline and isoquinoline and the ^{1}H data for phenanthrene [147]. In the second study [146], the exchange pattern was measured at an early stage at 14 per cent. to 46 per cent. approach to equilibrium. Under these conditions, scrambling of incorporated isotope does not occur (or at least is greatly reduced) and the ^{3}H distribution probably reflects true selectivity. The results show that when catalysis is carried out at 130°C, exchange in alkylpyridines occurs preferentially at a position adjacent to the ring nitrogen atom or at the 4-position (if not blocked). The effect of the methyl substituent(s) is complex in that a 2-methyl group inhibits exchange at the 3-position and to some extent at the 6-position, a 3-methyl group reduces exchange at the 4- and 5-positions but not at the 2- or 6-positions, whilst a 4-methyl group only inhibits exchange at the 3,5-positions. There is some analogy with alkylaromatic exchange: the nitrogen lone pair electrons in heterocyclic compounds increases the interaction between the substrate and catalyst surface, leading to a preference for the dissociative mechanism.

The previous examples have been concerned with measurement of the selectivity of different catalysts towards C–H bonds. Exchange in compounds such as organosilanes [148, 149] offers the opportunity of studying exchange at both C–H and Si–H bonds. Such studies have provided an easy route to tritiated tetramethylsilane and hexamethyldisiloxane which could be used as a ^{3}H internal reference [148], although the ghost referencing procedure is much to be preferred. It is significant in this work that no exchange using tritiated water has yet been achieved—in the few cases of successful exchange reported [147, 148] tritium gas was used as the source of the label and Raney nickel was the catalyst. Amongst the compounds successfully labelled are triethylsilane and chloro-dimethylsilane, the directly bonded Si–H hydrogen accounting for between 37 and 57 per cent. of the total incorporated tritium.

The ability of silicon to exceed four-coordination raises the possibility that species such as **XL** and **XLI** may exist on a metal surface [150]. The observation of Si–H exchange suggests that the dissociative σ-bonded species (**XLII** and **XLIII**) may also form on the surface of the Raney nickel. These are analogous to species participating in the exchange of alkanes on metal surfaces. The results raise the interesting possibility that primary adsorption of the organosilane may occur through the silicon followed by interconversion of species **XL** to **XLII** and **XLI** leading to exchange at the silyl position.

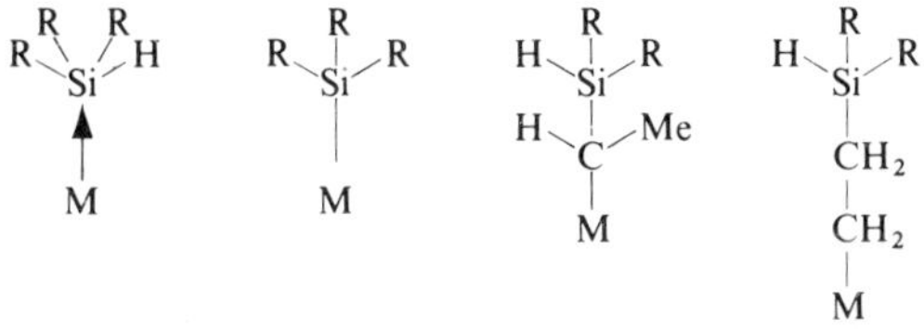

(XL) (XLI) (XLII) (XLIII)

Another aspect of the ^{3}H nmr studies of catalysis concerns the tritiation of long-chain esters [151] such as hexylpropionate. This has been achieved using tritium gas and a rhodium black catalyst. With the aid of an europium shift reagent (Eu(fod)$_3$) it was possible to resolve all the methylene signals and show that the labelling levels were highest at the two ends of the molecule and decreased steadily towards the ester grouping, the intensity of the signal from tritium on the methylene group next to the oxygen group in the hexyl group being close to zero. These results suggest that the extremities of the molecule may well be the favoured sites of attachment to the catalyst.

Silica gel, irradiated with γ-rays, has been used [144, 151] for the promotion of exchange between tritium gas and organic compounds. A ^{3}H nmr study of the orientation of tritium, introduced in this way into a variety of branched alkanes and their halo-derivatives, shows that incorporation takes place preferentially in the methyl groups adjacent to tertiary carbon atoms. In the case of alkylaromatics the pattern of substitution is essentially that expected of an electrophilic mechanism. A similar labelling pattern (Table 39) is witnessed [94] when zeolite catalysts are employed in the presence of a trace of high specific activity tritiated water at a temperature of 175°C. The procedure represents a highly efficient method of tritiation of most aromatic compounds. Only in the case of severely deactivated aromatics, such as α,α,α-trifluorotoluene or bulky compounds such as triphenylmethane, does exchange appear to be substantially hindered. Attempts to label *n*-alkanes were unsuccessful and in the case of branched-chain alkanes

Table 39. Distribution of tritium in compounds labelled by HY-zeolite catalysis.* (Reprinted with permission from *J. Am. Chem. Soc.*, **103**, p. 1572 (1981). Copyright 1981 American Chemical Society.)

Compound	Time (h)	^{3}H Incorporation (%)	^{3}H Distribution (% relative per site)			
			Ortho	Meta	Para	Alkyl
Ph–CH_3	7	100	29.5	3.3	34.4	< 1
Ph–C_7H_{15}–*n*	48	22	26.5	2.2	42.5	< 1
Ph–$CH(CH_3)_2$	15	43	7.5	< 1	16.2	†
Ph–$CH(CH_3)CH_2CH_3$	16	74	15.9	2.6	30.9	‡
Ph–Br	24	30	13.2	< 1	73.6	
Ph–NH_2	74	11	39.5	< 1	21.1	
Ph–OH	73	93	30.2	7.0	25.5	
Ph–OCH_3	2	65	32.2	< 1	35.6	< 1
			α	β		
Furan	74	62	50	< 1		
Thiophen	74	100	28	22		
Naphthalene	20	27	22	2.8		
Ph–$Si(CH_3)_2H$	40	7.6	33		33	§

* Zeolite prepared from Linde SK40, exchanged (NH_4NO_3) to 3 per cent residual Na, and activation at 470 °C.

† CH(< 1), CH_3(11.5).

‡ CH(< 1), CH_2(6.7), βCH_3(6.2), γCH_3(< 1).

§ SiH(< 1).

these were usually converted to a range of isomerization products, many of which contained tritium.

The above study shows that zeolites are active catalysts for promoting isotope exchange at temperatures considerably below that at which they are used commercially to activate hydrocarbon conversion processes. If the zeolites are loaded with catalytically active noble metals, exchange of elemental tritium readily occurs [152] with both aromatic and aliphatic hydrocarbons at even lower temperatures, e.g. 100°C. The tritium orientation in the labelled products shows characteristics of both metal and zeolite catalysis. Thus in the case of toluene (Table 40) the methyl exchange and 'ortho deactivation' are phenomena associated with catalysis by metals such as platinum and palladium, whilst ortho/para predominance is typical of H-zeolites [94]. With the metallic forms of ZSM-5 no alkyl exchange was observed, presumably reflecting the constraints imposed by the particular pore geometry of the zeolite. The hydrogen forms of the same zeolites (i.e. without metal) also activate elemental tritium for aromatic exchange, but at a slightly higher temperature (120 to 150°C).

Besides their use in the study of isotopic exchange, catalytic hydrogenation reactions are extensively used to prepare tritiated compounds at high specific activity. Again the characteristics of different kinds of catalysts can be examined through the use of ^{3}H nmr spectroscopy. Thus in the heterogenous catalytic hydrogenation of cinnamic acid [96] (Table 41) with a hydrogen–tritium mixture

Table 40. Distribution of tritium in toluene labelled by exchange with tritium gas over zeolites. (Reproduced by permission of CSIRO.)

Catalyst	Tritium incorporation (%)*	Tritium distribution (% relative per site)			
		Ortho	Meta	Para	Me
PtY†	67.8	14.6	6.0	42.0	5.6
Pt Mordenite	59.4	13.3	8.0	44.5	4.3
PtZSM-5	66.8	19.3	19.0	23.4	<1
PdY	60.7	3.0	<1	15.7	26.1
Pd Mordenite	66.2	10.3	2.6	19.0	18.4
PdZSM-5	62.8	20.6	19.0	20.8	<1
HY‡	10.6	24.0	15.5	20.9	<1
HZSM-5§	2.3	20.7	19.9	18.7	<1
H-Mordenite¶	1.1				

* Equilibrium represents ca. 100 per cent. incorporation.
† 72 h at 100 °C.
‡ 168 h at 150 °C.
§ 168 h at 130 °C.
¶ 173 h at 125 °C.

Table 41. Distribution of tritium in [^{3}H]dihydrocinnamic acid

Catalyst (5%)	Tritium (%)		
	C–2	C–3 (benzylic)	Other (aromatic)
Pd/C	50	50	
Pt/C	19.5	47	33.5
$Pd/CaCO_3$	62.5	37.5	
$Pd/BaSO_4$	59	41	
Rh/C	57.5	42.5	
Pt/Al_2O_3	25	37.5	37.5

in the ratio of 9:1, all catalysts apart from 5 per cent. Pd/C show unequal addition across the double bond. Use of Pt catalysts results in a larger amount of tritium being added at the benzylic C–3 position whereas the other catalysts favour C–2. The Pt catalysts also result in significant amounts of exchange into the aromatic ring.

In contrast with the results obtained for cinnamic acid using 5 per cent. Pd/C, addition of tritium to styrene (Table 42) with this catalyst leads to a very uneven addition of tritium across the double bond, the methyl methylene ratio for tritium content being about 85:15 as shown by the ^{3}H spectrum (Figure 53a). When the tritiation was performed using a smaller quantity of gas than would be required to fully saturate the double bond, additional signals appeared in the ^{3}H nmr spectrum (Figure 53b). Radio-gas chromatography clearly showed the formation

Table 42. Distribution of tritium in [^{3}H]ethylbenzene and [^{3}H]styrene

Catalyst (5%)*	Tritium distribution (relative %)				
	Ethylbenzene		Styrene		
	Methyl	Methylene	Cis	Trans	Aromatic
Pd/C	35	10	28	27	
Pt/C	54	26	14	6	
Pt/Al_2O_3	28	11	35	9	17
Rh/C	48	12	20	20	
Raney nickel	62	30	4	4	
Ru/C	37	9	26	28	

* Except Raney nickel.

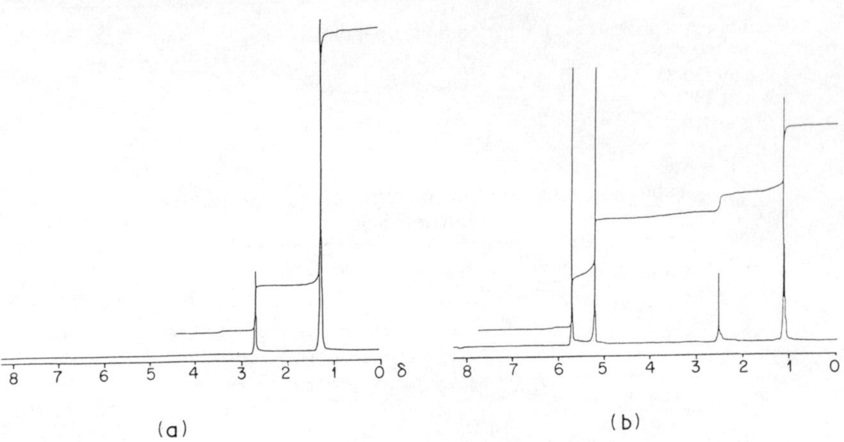

Figure 53. ^{3}H Nmr spectra (^{1}H-decoupled) of the reaction product (in $CDCl_3$) from addition of tritium to styrene in the presence of 5 per cent. Pd/C: (a) full reduction and (b) partial reduction, employing a deficiency of gas. (Reproduced by permission of The Royal Society of Chemistry.)

of two radioactive products, ethylbenzene and styrene. The additional ^{3}H nmr signals at $\delta = 5.2$ and 5.8 represented exchange at the terminal vinylic protons in styrene, trans and cis to the ring, respectively. This vinylic exchange occurs with all the catalysts but to a much lesser extent with Raney nickel, and it occurs to an equal extent at both cis and trans terminal positions with all the catalysts except the Pt ones which favour the cis-position. If the amount of gas employed in the hydrogenation is gradually increased to the theoretical and above, the ^{3}H nmr spectrum of the product mixture shows a gradual decrease of the signals due to exchange and a corresponding increase in the signals due to reduction. These observations are consistent with the involvement of the following processes:

$$RCH_a = CH_2 \xrightarrow{T_2} \begin{matrix} RCH_a = CHT \\ (+RCH_a = CT_2) \end{matrix} \xrightarrow{T_2} \begin{matrix} RCH_aT\text{–}CHT_2 \\ (+RCH_aT\text{–}CT_3) \end{matrix}$$

It is notable that exchange at the proton at the aryl-substituted end of the double bond, i.e. H_a, is not observed. Similar results to the above were also obtained for other substrates with comparable olefinic groupings, e.g. 4-vinylpyridine, 4-vinylcylohexene and 3-phenylpropene [87].

As mentioned previously (page 167), reduction of alprenolol (**XLIV**) to dihydroalprenolol gives rise to a product whose ^{3}H nmr spectrum (Figure 54a) shows incorporation of tritium at all three positions in the side chain, with a methyl methylene benzylic incorporation ratio of 9:3:1. With less than saturating amounts of the hydrogen–tritium gas mixture, the products gave a ^{3}H nmr spectrum (Figure 54b) indicating that vinylic exchange together with the known

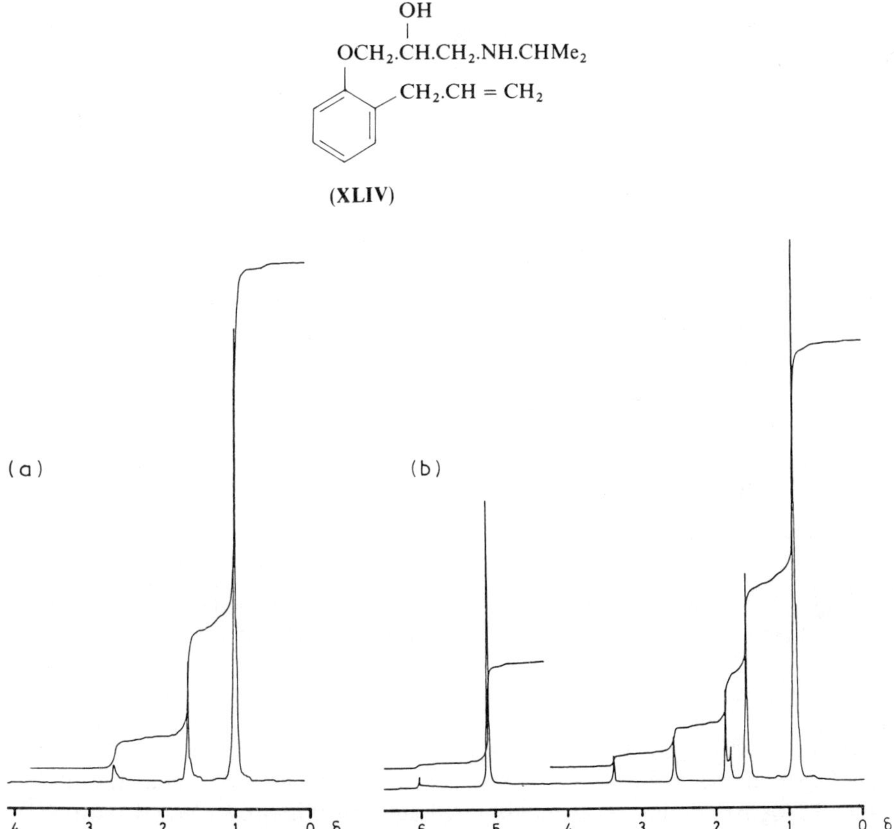

Figure 54. ^{3}H Nmr spectra (^{1}H-decoupled) of the reaction product (in $CDCl_3$) from addition of tritium to alprenolol (**XLIV**) with Pd/C (5 per cent.): (a) full reduction and (b) partial reduction. (Reproduced by permission of The Royal Society of Chemistry.)

processes of allylic and benzylic exchange, as well as double-bond migration, all contribute to the final labelling pattern. The results suggest that the following species arise:

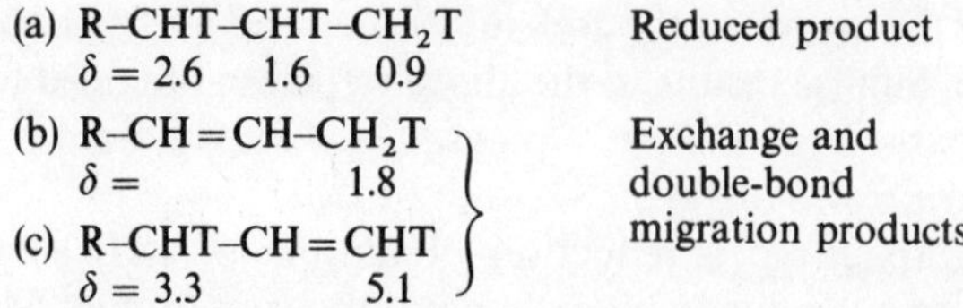

When a homogeneous catalyst such as Wilkinson's is employed in the hydrogenation of olefinic-type bonds the pattern of labelling is usually very similar to that obtained with heterogeneous catalysts. In some cases tritium addition occurs more evenly as in the case of dihydroalprenolol (Figure 55), but the most significant difference lies in the fact that double labelling occurs when using a 9:1 hydrogen–tritium mixture. This result implies that the exchange reaction $H_2 + T_2 \rightarrow 2HT$ is not as fast in these circumstances as clearly it is under homogeneous conditions.

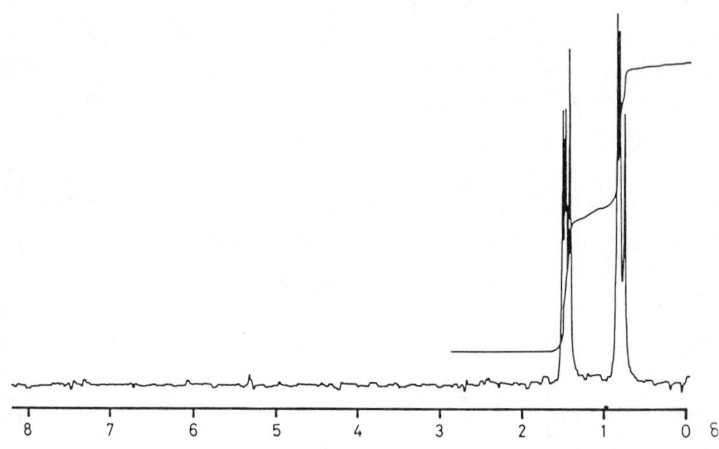

Figure 55. ^{3}H Nmr spectrum of the product (in $CDCl_3$) from reduction of alprenolol with tritium and homogeneous catalysis. (Reproduced by permission of The Royal Society of Chemistry.)

Hydrogenation using Wilkinson's catalyst can provide a simple route to amino acids containing a chiral methyl group [133]. Thus catalytic reduction of *N*-acetylisodehydrovaline by an equilibrated mixture of hydrogen and tritium (7:1) in the presence of Wilkinson's catalyst, followed by hydrolysis, gave a mixture of labelled diastereoisomers in which the (2*SR*, 3*SR*, 4*RS*)-[4-^{3}H,^{2}H,^{1}H]-*N*-acetylvaline predominated over the (2*SR*, 3*RS*, 4*SR*)-diasteroisomeride, as previously described (page 177).

In 1966 Wilkinson [153] discovered the highly active homogeneous rhodium catalysts for hydrogenation. Of these, tris-(triphenylphosphine)rhodium(I) chloride is now known as Wilkinson's catalyst. Since then many chiral diphosphines have been developed [154, 155] in the expectation that they will be efficient catalysts for asymmetric reductions. The great interest in highly tritiated peptides for biological investigations has prompted a study [77] of the stereospecific tritiation of the double bond in dehydropeptides such as (*Z*)-*N*-acetyldehydrophenylalanyl-(*S*)-phenylalanine methyl ester (**XLVI**). The catalysts most frequently used have the molecular composition L_2RhCl, where L_2 is a chiral diphosphine such as (+)diop (**XLVa**), (−)diop (**XLVb**) [156], dipamp (**XLVc**) [157] or bppm (**XLVd**) [158], or are the chelates $[L_2RhCOD]^+ X^-$ in which COD represents 1,5-cyclooctadiene.

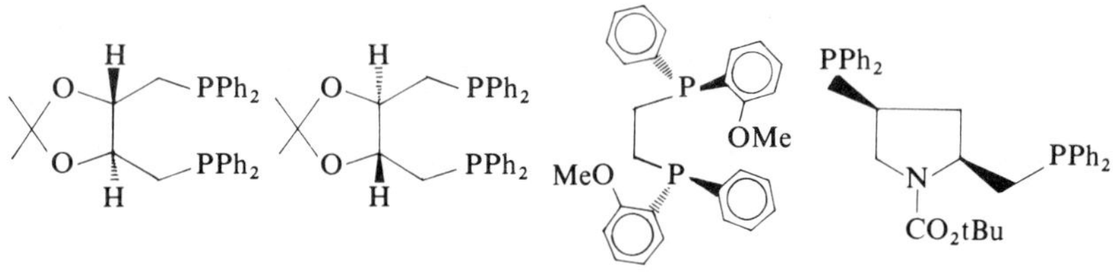

(+)diop (**XLVa**)

(−)diop (**XLVb**)

dipamp (**XLVc**)

bppm (**XLVd**)

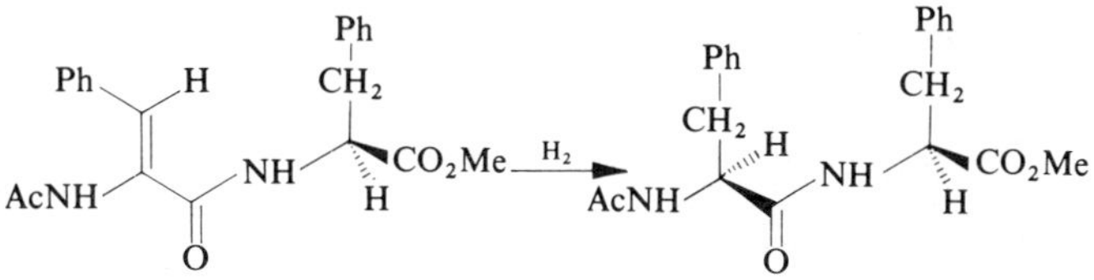

(*Z*)–(*S*) (**XLVI**)

(*S*, *S*) (**XLVIIa**)

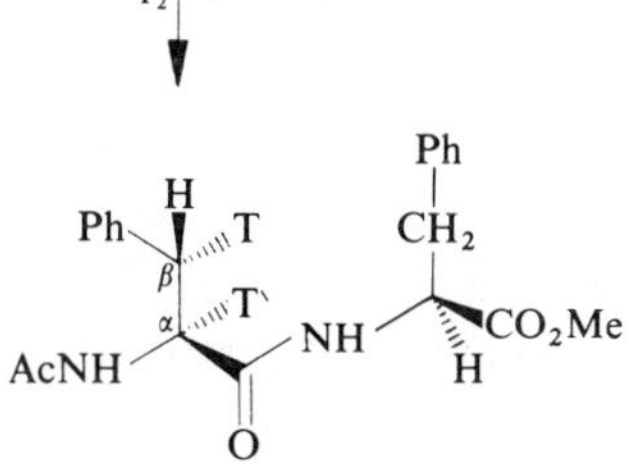

+

(*R*, *S*) (**XLVIIb**)

($S\beta_1$, $S\alpha_1$, S_2) (**XLVIIc**)

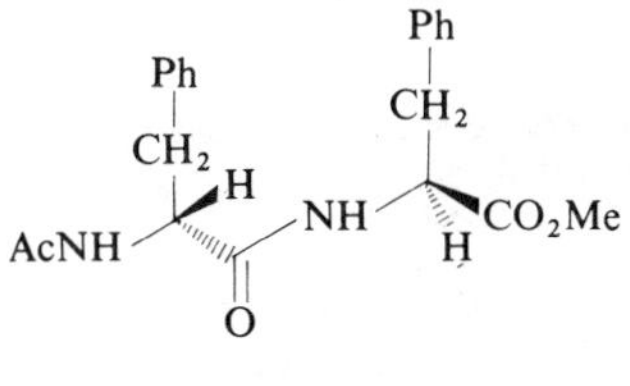

+

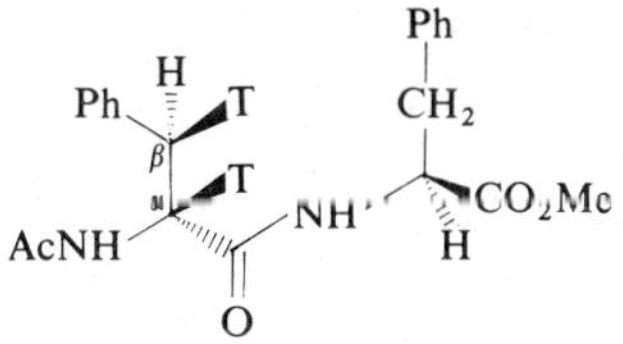

($R\beta_1$, $R\alpha_1$, S_2) (**XLVIId**)

When the dehydrophenylalanylphenylalanine precursor (**XLVI**) is reduced to **XLVIIa** + **XLVIIb** with hydrogen gas in the presence of the above catalysts, the following stereoselectivities were obtained [77]: (+)diop, (**XLVIIa**): (**XLVIIb**) = 95:5; (−)diop, 10:90; dipamp, ⩾95: ⩽5; bppm, ⩽5: ⩾95. In the corresponding reduction with tritium using (+)diop, **XLVI** was reduced with the high stereoselectivity, (**XLVIIc**):(**XLVIId**) = 88:12. The diasteroisomers were separated and then analysed by ^{3}H nmr spectroscopy (Figure 56). In compound **XLVIIc**, both α and β sites were identically labelled whereas in **XLVIId** the small deficiency at C_α could have arisen from isotope exchange with the solvent. The ^{1}H and the ^{3}H nmr spectra make it possible to eliminate the ambiguities of assignment of the β- and β'-protons in **XLVIIa** and **XLVIIb** and to calculate the fractional population of the three principal rotamers about the C_α–C_β bond:

(**XLVIIc**)

	Rotamer 1	Rotamer 2	Rotamer 3
Front (top)	CO	CO	CO
Back, upper left	H_{β}'	(T)H_β	Ph
Back, upper right	H_β(T)	Ph	H_{β}'
Front, lower left	AcHN	AcHN	AcHN
Front, lower right	H_α(T)	H_α(T)	H_α(T)
Back (bottom)	Ph	H_{β}'	H_β
^{1}H nmr	0.61	0.17	0.22
^{3}H nmr		0.18	

(**XLVIId**)

	Rotamer 1	Rotamer 2	Rotamer 3
Front (top)	NHAc	NHAc	NHAc
Back, upper left	Ph	H_{β}'	(T)H_β
Back, upper right	H_{β}'	H_β(T)	Ph
Front, lower left	OC	OC	OC
Front, lower right	H_α(T)	H_α(T)	H_α(T)
Back (bottom)	H_β(T)	Ph	H_{β}'
^{1}H nmr	0.20	0.67	0.13
^{3}H nmr			0.12

Given the very high stereoselectivity achieved, it should be possible to prepare other types of dipeptides or longer peptides at high optical purity. Indeed, the tritiation of locust adipokinetic hormone has similarly been effected. Thus catalytic reduction with tritium gas of the didehydropeptides having 4,5-didehydroleucyl and 3,4-didehydroprolyl residues gave samples of [^{3}H]hormone. That the labelling in the tritiated hormone from the first reduction was confined to the 4,5-leucyl sites was established by ^{3}H nmr spectroscopy [159].

In another recent report on tritium labelled peptides, high specific activity [^{3}H]acetylmuramyl peptides were prepared, either by using [^{3}H]acetic anhydride or by iodoacetylation with *N*-(iodoacetoxy)succinimide in chloroform followed by tritiodeiodination with tritium gas and a catalyst. ^{3}H Nmr spectroscopy showed that the latter product had 96.6 per cent. of the tritium present in the *N*-acetyl group and 1.9 per cent. at the hemiacetal position, with the remaining 1.5 per cent. in several other sites. ^{3}H Nmr spectroscopy also revealed that tritium introduced into *N*-acetylmuramyldipeptide, by catalysed exchange

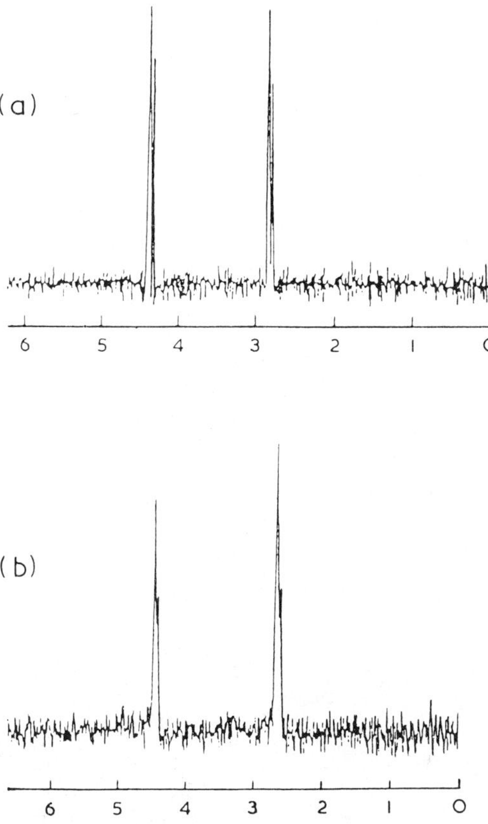

Figure 56. ^{3}H Nmr spectra for solutions in d_6-DMSO of (a) ($S\beta$,$S\alpha$)-and (b) ($R\beta$, $R\alpha$)-[α,β-^{3}H]N-acetylphenylalanyl-S-phenylalanine methyl ester, (**XLVIIc**) and (**XLVIId**) respectively. (Reprinted with permission from *Tetrahedron Lett.*, p. 4172 (1978). Copyright 1978, Pergamon Press.)

with tritium gas, resided in the *N*-acetyl and the hemiacetal positions to the extent of 64 per cent., with the rest extensively distributed [160].

4. Reaction mechanisms studies

The reaction of phenols with chloroform and aqueous alkali to give *o*- and *p*-hydroxybenzaldehydes, discovered by Reimer and investigated in collaboration with Tiemann, is now well known as the Reimer–Tiemann reaction. Subsequently, the mechanism was thought to be an example of an S_N2 type with attack on the *o*- or *p*-carbon atom of the phenoxide by chloroform, followed by tautomerism of the product to a substituted benzylidene dichloride that is then rapidly hydrolysed:

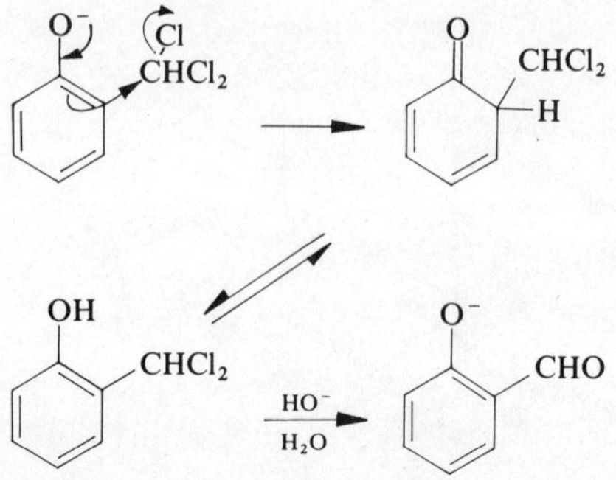

Hine [161], however, showed that chloroform reacts only very slowly with sodium phenoxide at 35 °C and that in the presence of sodium hydroxide the reaction is much faster. These results ruled out the above mechanism and suggested that firstly the hydroxide ions react with chloroform to give dichloromethylene which then reacts as an electrophile with the phenoxide ion:

$$CHCl_3 + \bar{O}H \longrightarrow \bar{C}Cl_3 + H_2O$$

$$\bar{C}Cl_3 \longrightarrow \bar{C}l + \ddot{C}Cl_2$$

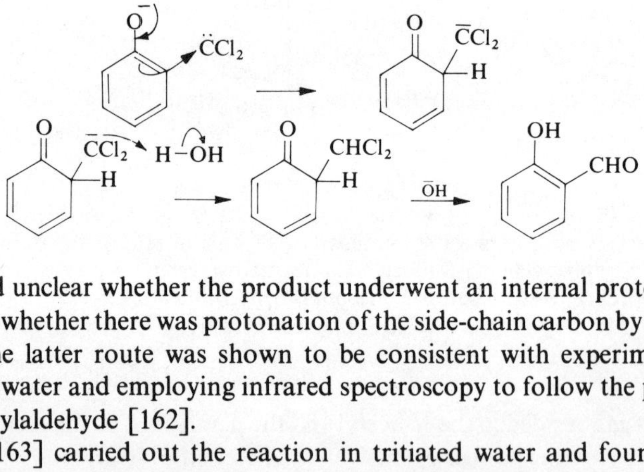

It remained unclear whether the product underwent an internal proton transfer reaction or whether there was protonation of the side-chain carbon by the solvent (water). The latter route was shown to be consistent with experiments using deuterated water and employing infrared spectroscopy to follow the production of the salicylaldehyde [162].

Kemp [163] carried out the reaction in tritiated water and found that the eventual product (salicylaldehyde) was radioactive. That virtually all of the radioactivity was located in the formyl group was demonstrated by means of the following reaction sequence:

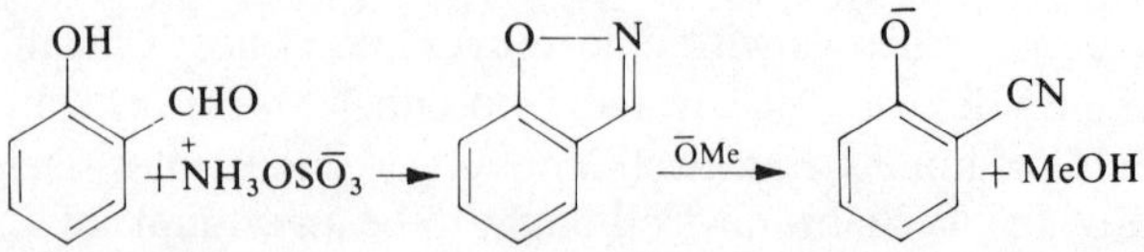

A far simpler procedure involved obtaining the ^3H nmr spectrum of the salicylaldehyde [5]. This showed a singlet, at 4.15 p.p.m. to low field of HTO as reference, characteristic of an aldehyde triton, and this finding together with the intensity of the signal served to confirm Kemp's results.

Results of a mechanistic nature emerged from a study directed at the preparation of α-labelled, α,β-unsaturated acids. For various acetic acids the method of Atkinson *et al.* [164] works very well. This involves the simple refluxing of the potassium salts of the acids in alkaline (e.g. 3 *M* KOH) tritiated water:

$$RCH_2COO^-K^+ \underset{\text{reflux}}{\overset{\text{KOH}}{\rightleftharpoons}} R\bar{C}HCOO^-K^+ \xrightarrow{\text{THO}} RCHTCOO^-K^+$$

Attempts to label cinnamic acid in this way, made knowing [109] that the α-positions of senecioic [164] and cinnamic acid [165] similarly exchange (at least under vigorous conditions), were, however, unsuccessful, but an alternative route proved satisfactory: this was the Knoevenagel condensation of benzaldehyde with malonic acid in the presence of base (primary or secondary amines), followed by decarboxylation in a tritiated water–dioxan mixture [109]. This approach was extended to the preparation of other [α-^{3}H]α,β-unsaturated acids. Also successful was the Doebner modification in which an arylaldehyde and tritiated malonic acid were heated together in boiling pyridine. When this procedure was extended to αβ-unsaturated aldehydes, e.g. crotonaldehyde and cinnamaldehyde, the products which were sorbic acid and δ-phenylsorbic acid, respectively, were found by ^{3}H nmr to be labelled at both α- and γ-positions (as previously mentioned on page 167). This indicated a mechanism involving lactonization [109]:

R–CH=CH–CHO + $CH_2(CO_2H)_2$ → R–CH=CH–CH=C(CO_2H)$_2$ → lactone (R, O, =O, CO_2H)

↓ γ-exchange

py: H, H– (lactone, R, O, =O, CO_2H) → R–CH=CH–CH=CH–CO_2H

Other mechanistic results came from a study of a route to tritium labelled taurine [76], one of many labelled amino acids that find wide application in the biochemical sciences. Its synthesis can be achieved by the following three-stage sequence:

$$\begin{matrix}CO_2Et \\ | \\ CH_2\overset{+}{N}H_3\overset{-}{Cl}\end{matrix} \xrightarrow[\text{2. HCl}]{\text{1. AlCl}_3/\text{NaB}\overset{*}{\text{H}}_4} \begin{matrix}C\overset{*}{H}_2OH \\ | \\ CH_2\overset{+}{N}H_3\overset{-}{Cl}\end{matrix}$$

$$\begin{matrix}C\overset{*}{H}_2OH \\ | \\ CH_2\overset{+}{N}H_3\overset{-}{Cl}\end{matrix} \xrightarrow[125\,^\circ C]{\text{HBr}} \begin{matrix}C\overset{*}{H}_2Br \\ | \\ CH_2\overset{+}{N}H_3\overset{-}{Br}\end{matrix}$$

$$\begin{matrix}C\overset{*}{H}_2Br \\ | \\ CH_2\overset{+}{N}H_3\overset{-}{Br}\end{matrix} \xrightarrow{Na_2SO_3} \begin{matrix}C\overset{*}{H}_2SO_3H \\ | \\ CH_2NH_2\end{matrix}$$

Perhaps not surprisingly, the ^{3}H nmr spectrum (^{1}H-decoupled) of the product in D_2O (Figure 57) showed two strong signals at $\delta = 3.22$ and 3.38, signifying the presence of tritium in both methylene groups. This could only have come about as a result of the formation of a symmetrical (aziridine) intermediate, an interpretation that was confirmed also by separate deuterium labelling studies [76]. The proposed mechanism therefore is as follows:

$$Na_2SO_3 + H_2O \rightleftharpoons NaOH + NaHSO_3$$

$$\begin{array}{l} \overset{*}{C}H_2Br \\ | \\ CH_2\overset{+}{N}H_3\overset{-}{Br} \end{array} + NaOH \longrightarrow \begin{array}{l} \overset{*}{C}H_2Br \\ | \\ CH_2NH_2 \end{array} + H_2O + NaBr$$

$$\begin{array}{l} \overset{*}{C}H_2Br \\ | \\ CH_2NH_2 \end{array} \rightleftharpoons \begin{array}{l} \overset{*}{C}H_2 \\ | \\ CH_2 \end{array}\!\!> \overset{+}{N}H_2\overset{-}{Br}$$

$$NaHSO_3 + \begin{array}{l} \overset{*}{C}H_2 \\ | \\ CH_2 \end{array}\!\!> \overset{+}{N}H_2\overset{-}{Br} \longrightarrow \begin{array}{l} \overset{*}{C}H_2SO_3H \\ | \\ CH_2NH_2 \end{array} + \begin{array}{l} CH_2SO_3H \\ | \\ \overset{*}{C}H_2NH_2 \end{array} + NaBr$$

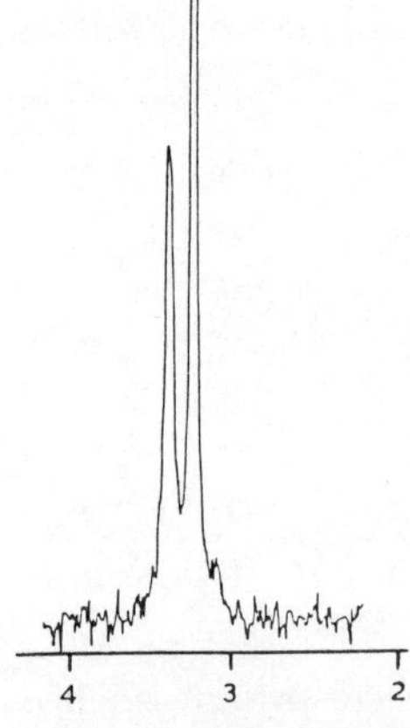

Figure 57. Proton-decoupled ^{3}H nmr spectrum of [G-^{3}H]taurine in D_2O. (Reproduced by permission of The Royal Society of Chemistry.)

Analogous observations were made during the synthesis of [1–^{13}C]taurine by a similar route. Here ^{13}C nmr spectroscopy showed scrambling of the label, indicating that the reaction proceeded partly (56 per cent.) through a [^{13}C]aziridine intermediate [166].

Base-catalysed hydrogen isotope exchange in nitroaromatics has been extensively investigated and several reviews [167 to 171], mainly dealing with Meisenheimer complex (or *J*-complex) formation, have appeared in the past decade. In the case of 1,3,5-trinitrobenzene the results are not always consistent but for 1,3-dinitrobenzene there is general agreement that exchange occurs only at the 2-position. This was recently confirmed by a ^{3}H nmr study [111] for the exchange in dioxan–methanol–water containing sodium methoxide. When a more basic system (hexamethylphosphoramide–methanol–water containing sodium methoxide) was employed, tritium was incorporated at C-4(6) as well as C-2, the relative amounts being 7 and 93 per cent., respectively. The positions tritiated could be easily identified by comparison of ^3H and ^1H nmr spectra (Figure 58). Kinetic studies at tracer level showed that the C-4 position was some 2,000 times less reactive than the C-2 position, and it is this large difference which probably explains why exchange at the C-4 position had not previously been observed.

The analysis of kinetic data from compounds containing two exchangeable sites is a straightforward matter when, as in the above case, the two rates differ substantially. There are, however, many instances where this may not be the case, e.g. in unsymmetrical ketones or in keto steroids where only the stereochemical configuration of the exchanging hydrogens is different. In such instances the reaction can be considered as consisting of two parallel steps. The relevant rate equation is:

$$\frac{N_0 - N_t}{N_0 - N_\infty} = x[1 - \exp(-k_{T(1)}t)] + (1 - x)[1 - \exp(-k_{T(2)}t)] \qquad (12)$$

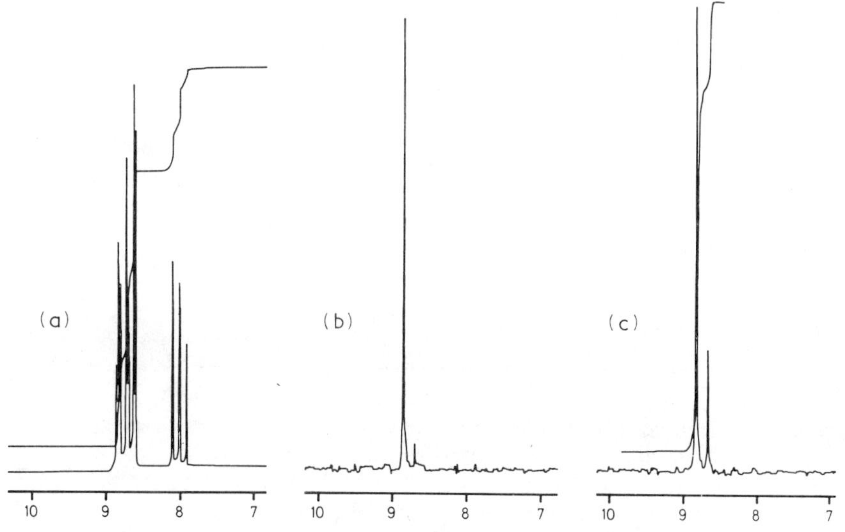

Figure 58. Nmr spectra of 1,3-dinitrobenzene in d_6-DMSO: (a) ^1H spectrum; (b) ^3H (^1H-decoupled) spectrum of product produced in dioxan–methanol–water system; (c) ^3H (^1H-decoupled) spectrum of product produced in hexamethylphosphoramide–methanol–water system. (Reproduced by permission of The Royal Society of Chemistry.)

where N_0, N_t and N_∞ are the substrate radioactivity (counts per minute) at times 0, t and ∞, respectively; x is the fraction of tritium at the more reactive site and $k_{T(1)}$ and $k_{T(2)}$ are the first-order detritiation rate constants from the most reactive and least reactive sites, respectively. Salomaa and coworkers [172 to 174] have made extensive use of this equation in their work on the base-catalysed detritiation of unsymmetrical ketones and substituted camphors; in the absence of ^{3}H nmr data they had to use a trial-and-error procedure in order to obtain values for x. A combination of tracer level kinetics and x values derived from ^{3}H nmr spectra provides an alternative approach to such problems.

When 1,3-dinitronaphthalene was subjected to base-catalysed hydrogen exchange [112] under conditions very similar to those used for 1,3-dinitrobenzene [111], the ^{3}H nmr spectrum (Figure 59b) showed that exchange had occurred at two positions, identified from the ^{1}H nmr spectrum (Figure 59a) as the 2- and 4-positions. However, in contrast to the observations with 1,3-dinitrobenzene, exchange at the 4-position had occurred to a greater extent (85 per cent.) than at the 2-position. Differences in carbenoid type delocalization, similar to those reported for heterocyclic systems [175], may be responsible for the observed differences.

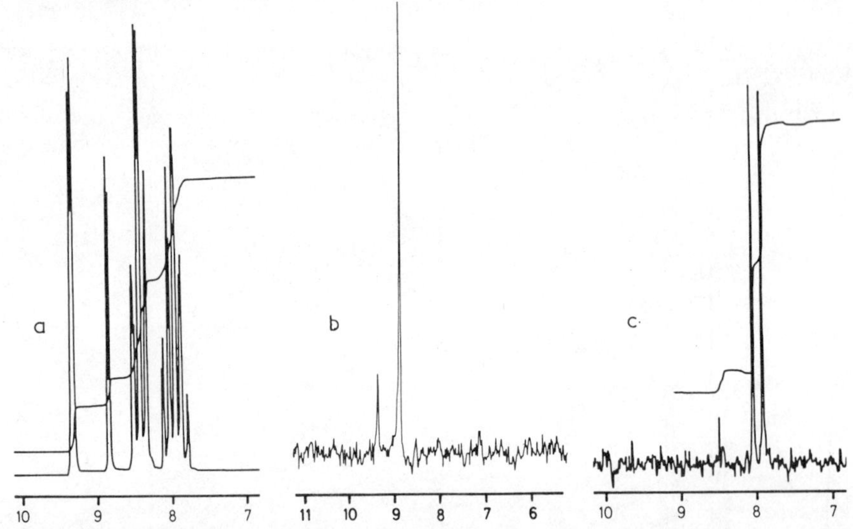

Figure 59. Nmr spectra of 1,3-dinitronaphthalene in d_6-acetone: (a) ^{1}H spectrum; (b) ^{3}H (^{1}H-decoupled) spectrum of base-catalysis product; (c) ^{3}H (^{1}H-decoupled) spectrum of product produced by tetrachloroplatinate catalysis. (Reproduced by permission of The Royal Society of Chemistry.)

When 1,3-dinitronaphthalene was subjected to exchange catalysed by potassium tetrachloroplatinate, 10 per cent. of the tritium was incorporated at C-5, with the rest equally distributed between C-6 and C-7 (Figure 59c). No exchange was observed at any of the positions close or adjacent to the nitro groups. These observations are in accord with the findings of Garnett and Hodges [176] on nitrobenzene, which pointed to a preference for exchange from positions furthest from the nitro group. These studies show that tritium labelling of 1,3-dinitronaphthalene can be achieved selectively, in either of the two rings, depending on the catalytic method employed.

5. Environmental chemistry

Whilst the uses of stable isotopically labelled compounds are generally unrestricted, necessary restrictions are imposed on the uses of radioactive compounds, which may therefore limit applications of compounds labelled with tritium for environmental investigations. Although low levels of tritium (as tritiated water) have been used, e.g. in numerous studies of hydrology [1], most applications of tritium labelled compounds for studies of the environment are confined to laboratory situations.

Tritium nuclear magnetic resonance spectroscopy has been used in two areas of environmental chemistry. The first of these is concerned with alternative energy resources and the second deals with aspects of health and safety.

Research designed to produce shale oil, as an alternative fossil fuel to augment dwindling petroleum reserves, is being actively pursued, particularly in the USA. Both surface and underground (in situ) oil shale processes are accompanied by the coproduction of process waters and a significant proportion of oil shale environmental research is being directed towards the characterization, transport, interaction, biological effects, treatment and utilization of these waters. The primary organic constituents include both polar and non-polar materials, with carboxylic acids and heterocyclic and aliphatic compounds being the important components. Because of the complex nature of these waters, studies based on a specific labelled component can provide little information, and the preferred approach would seem to be one in which the complex mixture is labelled. Such a labelled mixture should conform to the following criteria:

(a) It should be representative of the total dissolved organic material originally present in the sample.
(b) It should be uniformly labelled in stable positions for all components present.
(c) It should be of high specific activity.
(d) It should be minimally contaminated with artifacts induced by isolation procedures or autoradiolysis.
(e) Its preparation should be reproducible.

Research [177, 178] has shown that, for a series of process waters, the objectives can best be achieved by employing a catalytic hydrogen isotope exchange procedure with Raney nickel as the catalyst. Comparative ^{3}H and ^{1}H nmr spectroscopy (Figure 60) provides an analytical method for measuring the labelling pattern and labelling proportion, and in this way provides a means for defining the radiochemical character of a tritiated sample. The degree to which the chemical shift tracing of a ^{3}H nmr spectrum can be superimposed on that from the corresponding ^{1}H nmr spectrum of a tritiated sample provides a measure of the labelling pattern. Totally non-specific and random incorporation would give rise to similar spectra (differing only in the spin-coupling detail). The closeness of this fit would diminish as specificity of incorporation increased. Likewise, the degree of correspondence between the ^{3}H and ^{1}H nmr signal intensities provides a measure of the labelling proportion. The integration traces

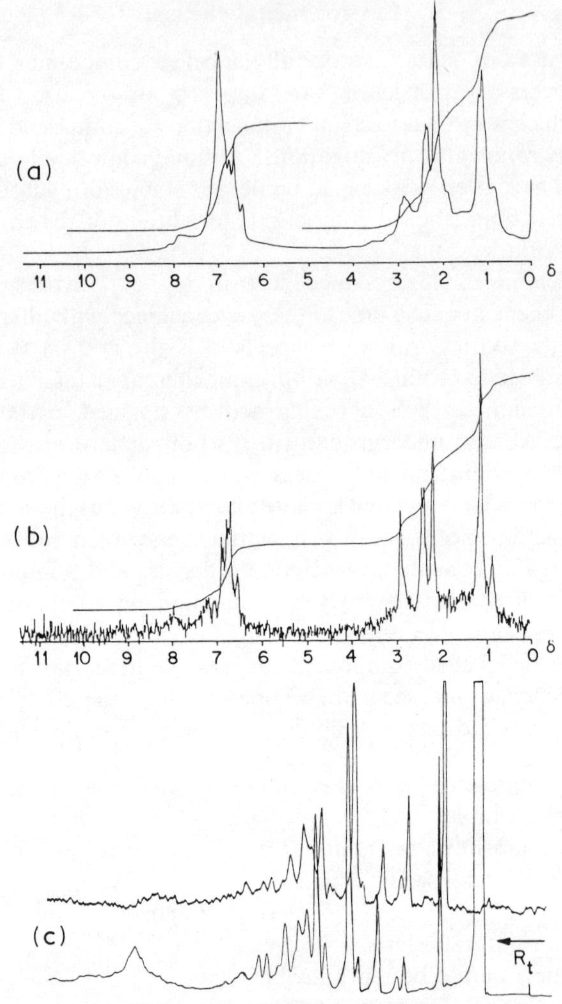

Figure 60. (a) ^{1}H Nmr spectrum of a complex organic mixture (in CD_3CN) derived from oil shale process water after exchange with tritiated water and Raney nickel; (b) corresponding ^{3}H nmr spectrum (^{1}H-decoupled); (c) gas chromatogram (detector response against retention time) and (upper trace) radio-gas chromatogram showing corresponding radioactivity against retention time. (Reproduced from W. P. Duncan and A. B. Susan, *Synthesis and Applications of Isotopically Labeled Compounds* by permission of Elsevier Science Publishers B. V.)

would give analogous quantitative results when the incorporated tritium is equally distributed in proportion to the protons available for exchange. The stability of the label can be readily checked by monitoring the ^{3}H nmr spectra of samples subjected to increased acidity or basicity, i.e. measured under more extreme conditions than those in which the labelled mixture is to be employed.

Radio-gas chromatography may also be used (Figure 60c) to substantiate the relationship between incorporated radioactivity and concentration of constituents.

Cancer is one of the most prevalent causes of death in the world. It can be induced by factors such as natural and synthetic carcinogenic compounds [179] and by radiation (X-rays, ultraviolet light and the emissions from radioactive isotopes). Some of these hazards can be avoided or reduced whilst others are an integral part of the environment in which we live. It is generally agreed that the cancer forming process, carcinogenesis, arises as a result of the interaction of the carcinogen with DNA, possibly by triggering the oncogenic portion of the DNA, but requires repeated potentiation. This viewpoint is supported by the known covalent binding of the metabolites of polycyclic aromatic hydrocarbons to DNA. The carcinogenic behaviour of polycyclic aromatic hydrocarbons appears to be a side-effect of the oxidation of these compounds, which probably proceeds either by epoxidation [180] at one or more double bonds, leading to a variety of dihydrodiols or through the production of quinones following hydroxylation [181].

In the past, one of the most effective ways of studying the binding of compounds to DNA was via the use of an appropriately labelled substrate. Thus Blackburn and coworkers [123, 182] have made extensive use of this method by first preparing the specifically tritiated aromatic hydrocarbon and observing how much label is lost on interaction with DNA; they then concluded that if the label is lost, binding is associated with chemical reaction at the labelled position. ^3H Nmr spectroscopy forms the basis of an alternative and attractive approach, especially as the amount of synthetic work can be greatly reduced by simply preparing a generally tritiated substrate using one of the many available one-step catalytic procedures. Thus a number of cyclopenta[a]phenanthrenes, as aromatic analogues of steroids, have been tritiated [110] using reduced PtO_2 catalyst. These compounds, which are so similar in chemical structure and physical properties, differ very markedly in biological activity. In the 17-ketone series the parent ketone, 15,16-dihydrocyclopenta[a]phenanthren-17-one (**XLVIII** with R = H) is inactive, the 11-methyl-17-ketone (**XLVIII** with R = CH_3) is a potent carcinogen similar in potency to benzo[a]pyrene, the 7-methyl-17-ketone is moderately active, whilst the 2-, 3-, 4-, 6- and 12-methyl-17-ketones are not carcinogenic. ^3H Nmr spectroscopic analysis of the generally tritiated ketone (**XLVIII** with R = CH_3) (Figure 61) showed that all positions are labelled with the exception of C-1, which is apparently sterically hindered by the 11-methyl

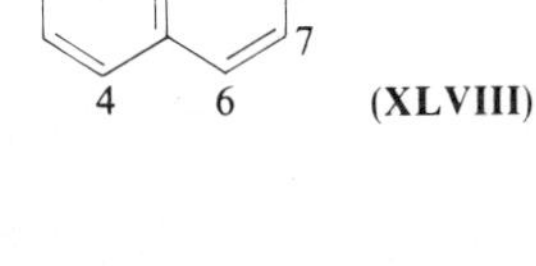

O
12
R 11
17
16
1
2
14
15
7
4
6

(**XLVIII**)

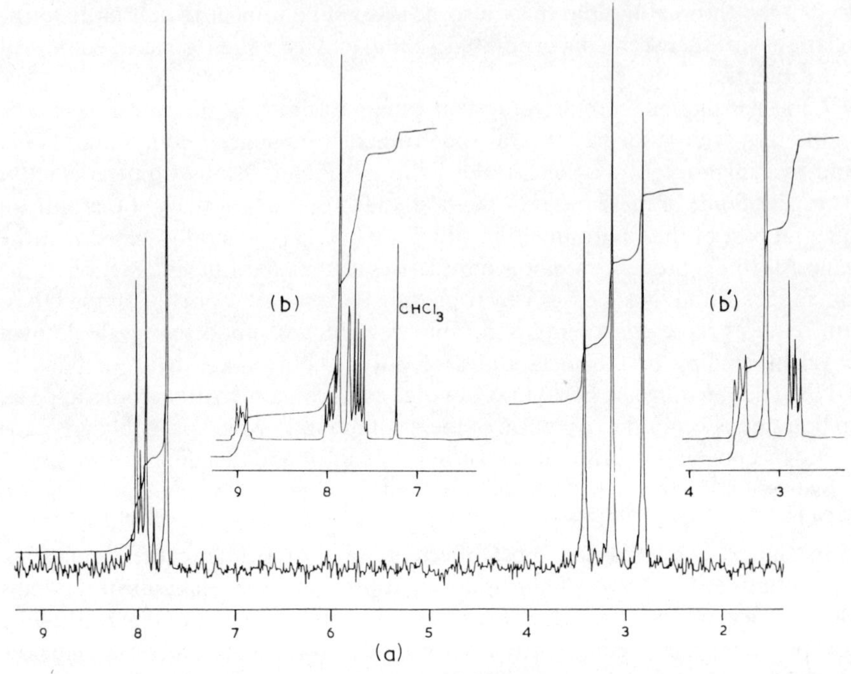

Figure 61. (a) ^{3}H Nmr spectrum (^{1}H-decoupled) of [G-^{3}H]15,16-dihydro-11-methylcyclopenta[a]phenanthren-17-one in $CDCl_3$, and (b,b′) ^{1}H nmr spectrum. (Reproduced by permission of John Wiley & Sons Ltd.)

group from approaching the catalyst surface. Also, labelling at the partially hindered C-7 and C-11 positions is lower than that at the unhindered C-2, -3, -4, and -6-positions; the 11-methyl group is particularly heavily labelled. This pattern is similar to that found for 3-methylcholanthrene and 7-methylbenz[a]anthracene labelled by the same procedure (see Chapter 2). Treatment of the compound with strong acid removed tritium at the α-position to the keto group quantitatively and tritium at benzylic hydrogen positions much less effectively.

Another application of ^{3}H nmr spectroscopy in cancer research concerns the preparation of 13-*cis*-[11-^{3}H]retinoic acid (**XLIXa**) [120]. The unlabelled compound has useful biological activity in preventing epithelial cancer in animals [120, 183]. For preparation of the labelled compound the reaction sequence starting from β-ionylideneacetaldehyde is similar to that used by Liebman and coworkers [184] for the synthesis of *trans*-[11-^{3}H]-α-retinoic acid. The tritiated β-ionylideneacetaldehyde (**L**) was prepared from the unlabelled aldehyde by reduction with sodium borotritide followed by reoxidation with manganese dioxide, with a net retention of over 75 per cent. of the tritium (a marked isotope effect). The phosphonium ylide (**LI**) used in the subsequent Wittig reaction was generated *in situ* from a mixture of *cis*- and *trans*-phosphonium chloride (**LII**) by sodium ethoxide in ethanol. Under these reaction conditions, the Wittig condensation of the tritiated aldehyde (**L**) with the ylide (**LI**) proceeded normally to give a mixture of four isomeric retinoic esters (**XLIXa** to **d**) of which the desired product (**XLIXa**) was the major component. Saponification of the ester with alcoholic potassium hydroxide readily afforded 13-*cis*-[11-^{3}H]retinoic acid. That this was labelled in the 11-position was shown by the singlet at $\delta = 6.88$ in the ^{1}H-decoupled ^{3}H nmr spectrum. The specificity of labelling in tritiated retinoic acid and retinol is important in studies of tissue distribution and subcellular localization and of cellular retinol binding protein(s) [183, 185, 186].

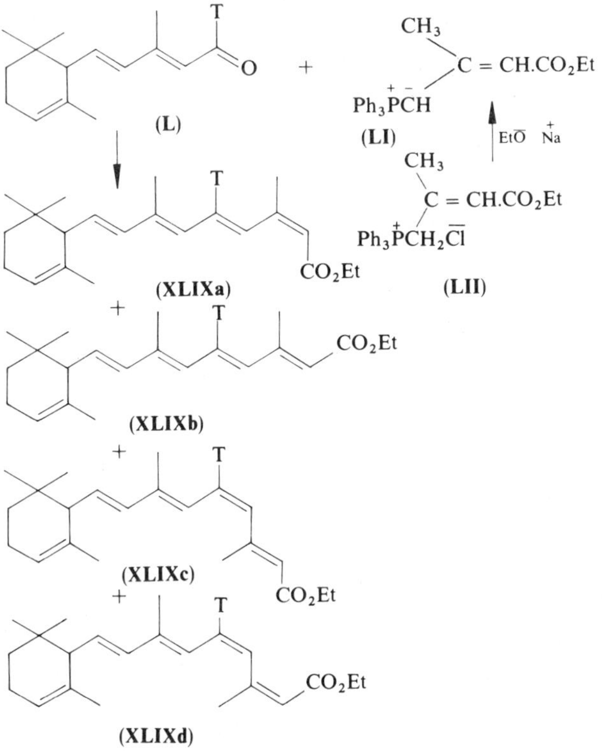

^{3}H Nmr spectroscopy is likely to find increasing use in the analysis of tritiated compounds used in studies of environmental problems.

6. Radiation chemistry

Tritium labelled compounds at very high specific activity tend to decompose as a result of self-radiolysis. For monitoring such changes, ^{3}H nmr spectoscopy should be ideal, as early indicated from observations [4] on [5-^{3}H]uridine (**LIII**) with the high specific radioactivity of 31 Ci mmol^{-1}. After 3 days, a 0.16 *M* solution of the compound had received ca. 9 megarads of radiation but the ^{3}H nmr spectrum was still normal, showing a sharp 1:1 doublet with ^3J(5T,6H) = 8.2 Hz, 1.22 p.p.m. downfield from tritiated water. After 6 days and 18 megarads of radiation, a signal had appeared at -2.00 p.p.m. (upfield from tritiated water) with $^2J\,(\mathrm{H, T}) \simeq 17$ Hz and $^3J\,(\mathrm{H, T}) \simeq 4.5$ Hz, together with a weaker doublet signal at -0.90 p.p.m. with $^3J\,(\mathrm{H, T}) \simeq 12.2$ Hz (Figure 62). These observations pointed to the formation of the hydrate (**LV**) and the glycol (**LIV**), respectively, in agreement with the findings concerning self-radiolysis of pyrimidines. This proceeds partly by the radiation-induced addition of water to the 5,6-double bonds [187] and partly by hydroxyl radical attack on the bond [1].

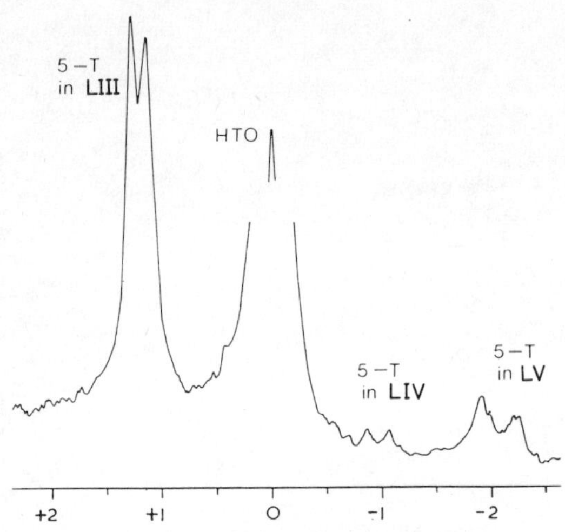

Figure 62. ^{3}H Nmr spectrum of [5-^{3}H]uridine (LIII) and its self-radiolysis products in water. (Reproduced by permission of John Wiley & Sons Ltd.)

After 5 weeks, corresponding to a radiation dose of ca. 100 megarads, most of the original [5-^{3}H]uridine had undergone radiolysis, as indicated by the virtual absence of the original low-field doublet. On the other hand, the highest field signal (at -2.0 p.p.m.) which arises from the hydrate (**LV**) was very strong, whilst

(**LIII**) R = β-D-ribose

(**LIV**) (**LV**)

Δ trans-(5T, 6H) → **LIII** + H_2O

cis-(5T, 6H) → uridine + HTO

that (at -0.9 p.p.m.) attributable to the glycol (**LIV**) remained roughly constant. By warming this solution, some of the hydrate (**LV**) was decomposed (the relevant signal at -2.00 p.p.m. diminishing) and the characteristic doublet signal (at $+1.2$ p.p.m.) from [5-^{3}H]uridine reappeared. The intensity was about 25 per cent. of that from the unchanged hydrate. The remaining signal at -2.00 p.p.m. was now much more clearly resolved as a double doublet, with splittings of 17.0 and 8.2 Hz, as would be expected from the *cis*-(5T,6H)-hydrate (**LV**). These several observations suggest that the *trans*-(5T,6H)-hydrate is preferentially decomposed to [5-^{3}H]uridine (**LIII**) and H_2O on heating.

DNA is extensively hydrated in solution and there is evidence [188, 189] to suggest that the hydration water molecules play an important part in mediating the radiation damage done to DNA by, for example, γ-rays. The hydration water can be studied specifically by nmr spectroscopy because a biological sample (or solution of macromolecules) in the frozen state contains the hydration water unfrozen. Only the nuclei of the latter will be moving sufficiently to give observable signals in a high-resolution nmr spectrum. There thus appears to be considerable potential for ^{3}H nmr spectroscopy in such studies, and some work [190] has already been published. Indeed, an understanding of the behaviour of tritiated water in the hydration layer of DNA is essential for describing the microdistribution of initial energy deposition when evaluating the radiobiological effectiveness of tritiated water in biological systems. The initial study shows that 3.5 per cent. of the tritiated water is present in the hydration layer. In view of the fact that the tritium β-rays are of low energy and therefore limited range (e.g. in water, maximum range 6.0 nm, mean range 0.69 nm) a highly localized radiation dose is received by the macromolecule.

Organic compounds containing two or more tritium atoms in the same molecule have been used in recent years in a variety of kinetic studies as the precursors of labelled carbocations of precisely defined structures [191 to 197]. A recent example [198], in which ^{3}H nmr spectroscopy was used, entailed the preparation of [1,4-$^3H_2$]benzene which in turn led to the radiation-induced formation of the phenylium ion (**LVI**). The resolution-enhanced proton-decoupled ^{3}H nmr spectrum (Figure 63) of the resultant benzene revealed the presence of two tritiated compounds, the most abundant (56 per cent.) being [1,4-$^3H_2$]benzene and the other (44 per cent.) [$^3H_1$]benzene. The chemical shift difference of 0.004 p.p.m. represents an isotope effect through five bonds, the first such reported.

T_2 $\xrightarrow{CuO}$ T_2O

Br, Br $\xrightarrow{Mg}$ MgBr, MgBr

T, T, +, T

(**LVI**)

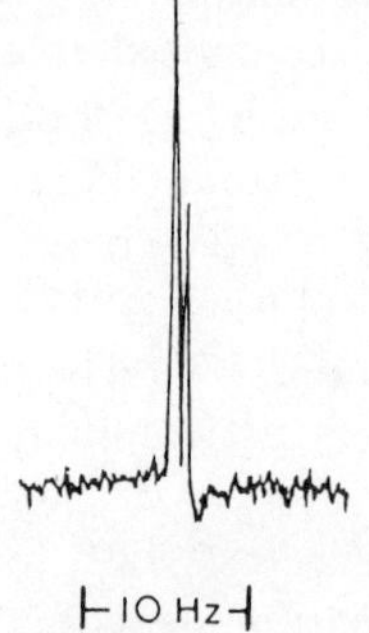

Figure 63. ^{3}H Nmr spectrum of the [$^{3}H_x$]benzene ($x = 1,2$) mixture acquired with resolution-enhancement and ^{1}H-decoupling. (Reprinted with permission from *J. Org. Chem.*, **45**, p. 3293 (1980). Copyright 1980 American Chemical Society.)

The high specific activity benzene (58 Ci mmol^{-1}) prepared in the above study undergoes rapid self-radiolytic decomposition when stored in the pure state. However a 200-fold dilution with inactive benzene in the presence of oxygen as a radical scavenger greatly reduces the effects of self-radiolysis, as the ^{3}H nmr spectrum then remains virtually unchanged during the course of 2 months.

In another study [199], multitritiated propane was prepared by passing propane with an excess of tritium gas over a nickel catalyst. The use of the product as a source of decay ions required a knowledge of the tritium distribution among non-equivalent molecular positions. The ^{3}H nmr spectrum showed two multiplets, characteristic of methylene and methyl groups, but because of the relatively low total activity of the sample the percentage abundance of the individual partially tritiated propanes could not be determined with any degree of confidence. Integration of the methylene and methyl multiplets gave a ratio of 2.4 ± 0.3, lower than expected for random labelling and indicative of preferential tritiation at the methylene positions.

7. Other analytical applications

The early demonstration [4] of the existence of a linear relationship between the intensity of the integrated ^{3}H nmr signal for tritiated water samples and the isotopic abundance as determined by liquid scintillation counting illustrated the potential of the method for obtaining the specific activity of a tritiated compound. This could be particularly useful for compounds which do not exhibit an ultraviolet absorption spectrum, as it is by such means that the very low concentration is usually determined. Likewise there could also be difficulties if the specific activity of the compound was very high and there were problems in accurately determining the mass, as was so with sodium borotritide [36]. The preparation was expected to produce $NaBT_4$, $NaBT_3H$, $NaBT_2H_2$ and $NaBTH_3$. As there are two boron isotopes (^{10}B and ^{11}B), there will be a total of 10 isotopomers including unlabelled species. Relevant nmr data are given in Table 43. The 200 MHz ^{1}H nmr spectrum of sodium borohydride in basic CD_3OD (Figure 64) consists of a quartet, centered at $\delta = -0.168$ p.p.m., due to coupling to the dominant ^{11}B isotope ($I = 3/2$) with $J = 80.6$ Hz, and a septet at

Table 43. Nmr spectral parameters of sodium borohydrides and -tritides.*

Isotopomer	Chemical shift (nuclear)	$^1J(^{11}B,H)$	$^1J(^{10}B,H)$	$^1J(^{11}B,T)$	$^1J(^{10}B,T)$	$^2J(H,T)$
$Na^{11}BH_4$	$-0.168(^1H)$	80.57				
$Na^{10}BH_4$	$-0.166(^1H)$		26.98			
$Na^{11}BH_3T$	$-0.125(^3H)$	80.1		86.4		11.14
$Na^{10}BH_3T$	$\sim -0.12(^3H)$					
$Na^{11}BH_2T_2$	$-0.149(^3H)$	79.5		85.6		11.13
$Na^{10}BH_2T_2$	$-0.144(^3H)$				28.7	
$Na^{11}BHT_3$	$-0.172(^3H)$	79.0		85.1		11.15
$Na^{10}BHT_3$	$-0.166(^3H)$				28.5	
$Na^{11}BT_4$	$-0.196(^3H)$			84.5		
$Na^{10}BT_4$	$\sim -0.19(^3H)$				28.3	

* Chemical shifts in p.p.m. (for ^1H the reference is TMS whereas for ^3H the internal TMS is converted to the ^3H ghost reference).

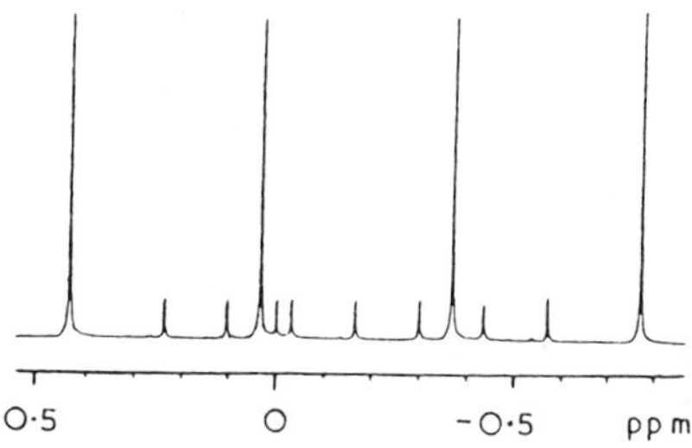

Figure 64. ^1H Nmr spectrum of sodium borohydride in d_4-methanol at 200 MHz. (Reprinted with permission from *Anal. Chem.*, **52,** p. 993 (1980). Copyright 1980 American Chemical Society.)

$\delta = -0.166$ p.p.m. with $J = 26.98$ Hz due to coupling to the minor ^{10}B isotope ($I = 3$). The corresponding ^3H nmr spectrum (Figure 65), obtained at 213.4 MHz and with proton spin decoupling, shows four quartets, one for each isotopically distinct species ($Na^{11}BH_{4-n}T_n$) as well as two septets, one each for $Na^{10}BH_2T_2$ and $Na^{10}BHT_3$: some of the septet lines from the other two tritiated [^{10}B]species are hidden by overlap with the main quartets. These assignments were confirmed from the proton-coupled ^3H nmr spectrum. Since each isotopomer gives rise to separate nmr signals it is possible to determine the mole per cent. of each present. The ^1H nmr spectrum gives the amount of unlabelled borohydride present and, if we assume that differences in (a) the spin lattice relaxation times and (b) the nuclear Overhauser enhancements for each isotopomer can be neglected, then the specific radioactivity can be calculated.

Table 44 shows a comparison of the molar specific activity as determined by ^3H nmr spectroscopy with direct measurement by liquid scintillation counting for a number of [*methyl*-^3H]compounds.

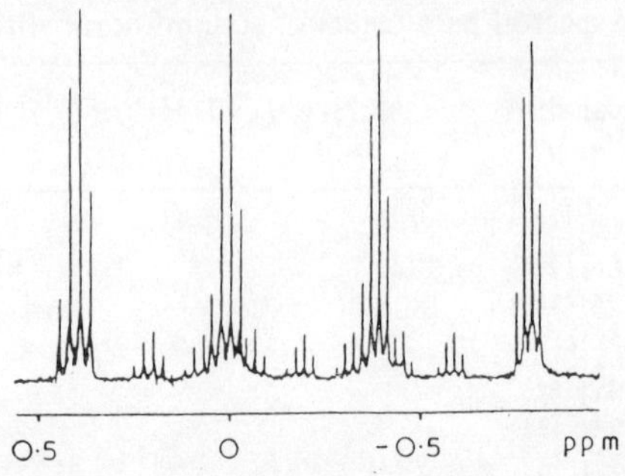

Figure 65. ^{3}H Nmr spectrum (^{1}H-decoupled) of sodium [$^3H_{1-4}$]borohydride in d_4-methanol at 213.47 MHz. (Reprinted with permission from *Anal. Chem.*, **52**, p. 993 (1980). Copyright 1980 American Chemical Society.)

Table 44. Specific activity of [*methyl*-^{3}H]compounds by ^{3}H nmr spectroscopy and by direct counting of radioactivity [19].

Compound	Specific activity ($Ci\ mmol^{-1}$) Scintillation counting	Tritium nmr spectroscopy
[*methyl*-^{3}H]Diazepam	84	85
[*methyl*-^{3}H]Flunitrazepam	84	85
[*methyl*-^{3}H]Methionine	81	85
[*methyl*-^{3}H]Methyltriphenyl phosphonium iodide	45	45
[*methyl*-^{3}H]Thymine	58	62

Specific activity measurements from ^{3}H nmr spectroscopic data have also been compared with those from mass spectrometry for a number of tritiated steroids labelled in the 16-position [106]. The data are summarized in Table 45. The specific activity by ^{3}H nmr spectroscopy was calculated as follows:

$$\text{Specific activity} = 57.52\,[1-(2R+1)^{-1}] \tag{13}$$

where 57.52 ($= 2 \times 28.76\ Ci\ mmol^{-1}$) is the currently accepted value for the maximum specific activity for the replacement of two hydrogens by tritium atoms, and R is the ratio between the mono- and ditritiated species from the ^{3}H nmr spectrum.

In general the specific activity determined from the ^{3}H nmr data for the steroids is lower than the results from mass spectrometry. This might be due to differential nuclear Overhauser effects, produced by ^{1}H-decoupling and favouring the monotritiated species, thus lowering the apparent $^3H_2/^3H_1$ ratio. Alternatively there might well be some isotopic ion preference in the mass spectrometric method.

Table 45. Specific activity from ^{3}H nmr and mass spectrometry. (Reproduced by permission of John Wiley & Sons Ltd.)

Compound	Ratio (R) $^3H_2/^3H_1$	Specific activity* (Ci mmol^{-1}) ^{3}H Nmr	Specific activity* (Ci mmol^{-1}) Mass spectrometry
[16-^{3}H]Desogestrel	0.27	20.2	22.7
3-Oxo[16-^{3}H]desogestrel	0.26	19.7	24.7
[16-^{3}H]Norgestrel	0.09	8.8	8.4
7α-Methyl[16-^{3}H]norethinodrel	0.08	7.9	8.0

* Data recalculated from that published [106] to correct for the currently accepted value of the half-life, and consequently the maximum specific activity, for tritium.

Reference has already been made to the use of ^{3}H nmr in obtaining inaccessible ^{1}H, ^{1}H spin–spin coupling constants [14] as well as the use to which chemical shift and coupling constant isotope effects can be put in the analysis of multitritiated species [19]. Another application of ^{3}H nmr spectroscopy is in the measurement of nuclear Overhauser effects [29, 32, 70] for ^{3}H nuclei as a result of ^{1}H-irradiation (noise decoupling).

Nuclear magnetic resonance spectra of partially oriented molecules in liquid crystal solvents can provide information on the molecular structure and ordering of the molecules [200]. A study [201] of benzene and several deuterated benzenes has demonstrated the existence of isotope effects on the degree of ordering and has shown that the direction of the isotope effect can test the models which describe the ordering process. In the first ^{3}H nmr study of molecules partially oriented in a liquid crystal, Wong and Altman [202] have analysed the ^{1}H and ^{3}H nmr spectra of a mixture of benzene and [$^3H_1$]benzene. The results show that the isotope effect is more evident on the degree of order than on the molecular structure. The best way to obtain the former is to compare directly the ratios of dipolar couplings in benzene and [$^3H_1$]benzene. The value of 1.0298 ± 0.0005 obtained is greater than that for [$^2H_1$]benzene but less than that for [1,4-$^2H_2$]benzene and consistent with the theory on ordering processes based on the predominant importance of dispersion forces. The experimental difference for C–H and C–T bond lengths obtained from this study (0.001 ± 0.003 Å) is small compared with the experimental uncertainties but in good agreement with the calculations based on the vibrational force field.

Solvent isotope effects are widely used in the study of reaction mechanisms and chemical equilibria [203]. They arise as a result of (a) an exchange effect, which represents the fact that hydrogen and, for example, deuterium are not randomly distributed between the various species present in a solution containing both isotopes, and (b) a transfer or medium effect, which arises from changes in solvation in going from one solvent to another. A key parameter in the interpretation of such isotope effects, as would arise in the following equilibrium:

$$SH + \tfrac{1}{2} D_2O \rightleftharpoons SD + \tfrac{1}{2} H_2O \tag{14}$$

is the fractionation factor ϕ which can be defined by

$$\phi = \frac{[SD]}{[SH]} \times \frac{[\tfrac{1}{2} H_2O]}{[\tfrac{1}{2} D_2O]} = \frac{(D/H)_{SL}}{(D/H)_{L_2O}} = \frac{F_{SL}}{1 - F_{SL}} \times \frac{1-n}{n} \tag{15}$$

where SH is the substrate with a labile hydrogen, L is a general hydrogen nucleus, F_{SL} is the fractional abundance of deuterium in the substrate and n that in water. If the substrate, or ion, bears no covalently bound exchangeable hydrogen as in the case of the methoxide ion in methanol the fractionation factor measures the ratio of concentration of MeOD to MeOH molecules at a solvation site relative to the bulk solvent [204].

Nuclear magnetic resonance studies on isotopically mixed solutions enable fractionation factors to be obtained directly. In principle one observes the chemical shifts in the nmr absorption of the hydroxylic protons (relative to some standard) as a function of solute concentration, firstly, in an entirely light (^1H) solvent and secondly in a partially deuterated solvent. Thus in the case of sodium methoxide solutions in methanol and MeOH–MeOD the fractionation factor ϕ_{OMe^-} is given by

$$\frac{S_n - \delta_0}{S_H - \delta_0} = (1 - n + n\phi_{OMe^-})^{-1} \tag{16}$$

where S_H and S_n are the limiting slopes and δ_0 contains contributions to the chemical shift from exchangeable hydrogens not subject to fractionation. The value of δ_0 is usually estimated by assuming that, in NaOMe, fractionation is confined to the solvation shell of the methoxide ion and that contributions to δ_0 arise only from the sodium ion. This is not entirely satisfactory and a simple way of resolving the difficulty and providing a more satisfactory evaluation of ϕ_{OMe^-} is to combine ^1H with ^3H nmr measurements [204]. In this way the number of measurements from which chemical shifts and fractionation factors may be determined are doubled without correspondingly increasing the number of parameters. The ^3H nmr study led to a revised value of $\phi_{OMe^-} = 0.7$ which differs but marginally from earlier values.

Isotope effects on chemical shifts have, as we have seen, analytical applications (pages 11, 73, 210). Although in general the effects are small there are some circumstances when they are appreciable. This is particularly true when hydrogen or one of its isotopes forms part of a hydrogen bond. In these circumstances isotope effects $\Delta\delta$ (^1H, ^2H) as large as 0.6 p.p.m. have been reported [205, 206], and these can be used to provide information relating to the shape of the potential energy surface in the vicinity of the equilibrium position. This arises from the fact that differences in the mass of the hydrogen isotopes influence the vibrational motions and the zero-point vibrational energies. The magnitude of the isotope effect is governed by the vibrational wave function ψ for the ground state and the variation of the chemical shift δ with the nuclear configuration. As regards chemical shift, the proton becomes deshielded as it moves towards the midpoint between the heavy atoms (usually O, N or F) involved in the hydrogen bond.

At least three types of potential minima may be encountered in hydrogen-bonded systems. In the case of weak hydrogen bonds (Figure 66a), the two potential minima are deep with a low anharmonicity approaching that of an ordinary covalent O–H bond. The effective equilibrium positions of the hydrogen isotopes (R) will then be clearly similar for ^{1}H, ^{2}H and ^{3}H, in which case the isotope effect will be close to zero. Such situations should be encountered in, for example, water or alcohols (see Table 46, compounds 1 to 3). For stronger and shorter hydrogen bonds (Figure 66b), the potential minima draw closer together with an accompanying decrease in the central barrier and increase in the anharmonicity of the potential. The equilibrium positions will now be different—$R_H > R_D > R_T$ (assuming R to be measured from the nearest heavy atom of the hydrogen bond)—and positive isotope effects in the chemical shift can be expected, larger for ^{3}H than for ^{2}H (Table 46, compounds 4 to 8). In the case of extremely strong and short hydrogen bonds (Figure 66c) the potential is symmetrical and the equilibrium distances become equal. However the difference in vibrational amplitudes between ^{1}H, ^{2}H and ^{3}H tend to make the chemical shifts of the isotopes increase with their mass, and because the heavier nuclei have

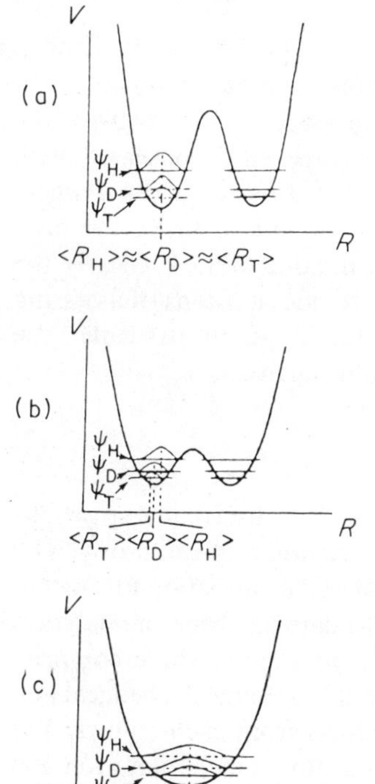

Figure 66. Relative positions of the hydrogen isotopes in the zero point vibrational levels for different hydrogen bond potential functions: (a) a double minimum potential with a high central barrier and low anharmonicity at the potential minima; (b) a double minimum potential with a low central barrier and high anharmonicity at the potential minima; (c) a single minimum potential. (Reprinted with permission from *J. Am. Chem. Soc.*, **100**, p. 8264 (1978). Copyright 1978 American Chemical Society.)

Table 46. Deuterium and tritium isotope effects on chemical shifts for hydrogen-bonding systems. (Reprinted with permission from *J. Am. Chem. Soc.*, **100,** p. 8625 (1978). Copyright 1978 American Chemical Society.)

Compound	$\delta(^1H)$ (p.p.m.)	$\Delta\delta(^1H,^2H)$ (p.p.m.) (± 0.03)	$\Delta\delta(^1H,^3H)$ (p.p.m.) (± 0.01)	$\Delta\delta(^1H,^3H$ / $\Delta\delta(^1H,^2H)$
1. Benzyl alcohol	2.00	−0.02	−0.016	
2. *o*-Hydroxyacetophenone	12.28	0.06	0.074	
3. Salicylaldehyde	11.02	0.06	0.008	
4. Ethyl acetoacetate	12.70	0.16	0.214	1.34 ± 0.28
5. Acetylacetone	16.11	0.61	0.830	1.36 ± 0.07
6. Benzoylacetone	16.59	0.67	0.932	1.39 ± 0.07
7. Dibenzoylmethane	17.61	0.72	1.071	1.40 ± 0.07
8. 1-Dimethylamino-naphthalene-8-dimethyl-ammonium cation	18.46	0.66	0.915	1.39 ± 0.07
9. Hydrogen maleate anion	20.5	−0.03	−0.07	
10. Hydrogen phthalate anion	21.0	−0.15	−0.25	1.66 ± 0.33

lower vibrational amplitudes a negative isotope effect should be observed (see Table 46, compounds 9 and 10).

Table 46 lists the deuterium and tritium isotope effects [207] on the hydrogen chemical shift measured under similar conditions for a number of systems with strong intramolecular hydrogen bonds. It is clear that the correlation between the hydrogen bond potential and the sign of the isotope effect observed with deuterium is substantiated by the ^3H measurements. For those cases where the isotope effect is positive and appreciable (compounds 4 to 8), the ratio $\Delta\delta$ (^1H, ^3H)/$\Delta\delta$ (^1H, ^2H) is close to the theoretical value of 1.44. Clearly the measurement of isotope effects in chemical shifts provides a means of assessing the shape of hydrogen bond potential energy functions, in particular the discrimination between double and single minimum potentials.

CONCLUDING REMARKS

This text has been written to review the first decade and a half of investigations into the technique of tritium nuclear magnetic resonance spectroscopy. The intention is to provide a firm foundation of fundamental information for further developments. The major application of ^3H nmr to date has been in analytical and structural investigations of tritium labelled compounds. These comprise mainly starting materials for research in the medical, biochemical, chemical and environmental sciences, but also the labelled products from such studies. The method is arguably the best for delineating the positions of tritium atoms in molecules, for giving the relative proportions of tritium in each of several sites simultaneously and for providing direct stereochemical information about the tritium label. The method is relatively very rapid and is non-destructive. Although lacking in absolute sensitivity as compared with the long established methods for detecting the presence of radioactivity, the nmr method has such important capabilities that the necessary use of millicuries amounts of tracer (rather than microcuries or lesser amounts) may often be warranted. Fortunately tritium is the least toxic and one of the least expensive radionuclides, and the necessary precautions in using millicurie amounts can be readily developed so as to be both simple and safe. Moreover, these precautions are easily reduced to a simple and safe routine, as has been outlined. There is no pre-requirement for the nmr spectrometer to be sited in a radiochemical laboratory or similarly protected area because the methods for sample handling are inherently very safe and designed to obviate contamination of the laboratory, even in the extremely unlikely event of the sealed sample being broken. The advantages of the nmr method over the conventional procedures of stepwise chemical degradation and counting are so great in terms of time, directness and certainty that the traditional approach is outmoded. In quantitative ^{3}H nmr spectroscopy there is generally no need to use the time-consuming special pulse sequence for full suppression of nuclear Overhauser effects. Differential effects are small in the average partially tritiated sample subjected to proton noise decoupling. Hence the straightforward integration of the relatively rapidly acquired ^{1}H-decoupled ^{3}H nmr spectrum gives acceptable results, these being at least as accurate as those from the tedious method of degradation and counting and free from the uncertainties of this latter approach. For the measurement of very high specific radioactivities, the ^{3}H nmr method will be the most precise available, whilst for the detailed analysis of multiply tritiated groups such as methyl or borotritide, the ^{3}H nmr method is unique. A field of investigation for which ^{3}H nmr could prove extremely helpful, as preliminary studies indicate, is that of the self-radiolysis of highly tritiated compounds. Other areas in which initial study has been made but which await further exploration by means of tritium labelling and ^{3}H nmr spectroscopy include the study of hydrogen bonding, of water of association and hydration, and of fractionation factors, investigations of highly complex mixtures, catalysis, and of rearrangement reactions in general and the reforming of alkanes upon zeolite catalysts in particular. In addition there are purely nmr spectroscopy problems to be solved.

In stressing the utility of the tritium isotope of hydrogen in this text, it is only fair to comment that the stable isotope deuterium has complementary uses. The deuteron, being a quadrupolar nucleus, has nuclear magnetic properties appropriate for the detailed study of molecular motion and orientation, as in investigations of biological membranes, micelles, etc. For problems soluble by simple chemical shift and intensity measurements, deuterium labelling at appropriately high abundance and either ^{1}H or ^{2}H nmr (the latter best at a very high field) will suffice. Where spectral dispersion is all important, e.g. when chemical shifts are very close and particularly when stereochemical problems need to be resolved through the measurement of short range and long range coupling constants, then it is virtually essential to use tritium as the tracer and

^{3}H nmr spectroscopy as the method of observation. The use of tritium does have the advantage that, when necessary, the established very sensitive techniques of autoradiography, scintillation counting, reverse isotope dilution analysis and radio-gas chromatography can also be applied to aid the solution of problems in which tritium labelling and tritium nuclear magnetic resonance spectroscopy play a major, essential part.

APPENDIX 1

International system of units (SI units) for radioactivity

The SI unit for radioactivity is the becquerel (Bq), equal to one disintegration per second. The table below gives some conversion values for becquerels to curies, and vice versa:

1 becquerel		1 Bq	27.03 picocuries
1 kilobecquerel	(1 kBq)	10^3 Bq	27.03 nanocuries
1 megabecquerel	(1 MBq)	10^6 Bq	27.03 microcuries
1 gigabecquerel	(1 GBq)	10^9 Bq	27.03 millicuries
1 terabecquerel	(1 TBq)	10^{12} Bq	27.03 curies
1 petabecquerel	(1 PBq)	10^{15} Bq	27.03 kilocuries
1 exabecquerel	(1 EBq)	10^{18} Bq	27.03 megacuries
1 microcurie	(1 μCi)	=	37 kBq
1 millicurie	(1 mCi)	=	37 MBq
1 curie	(1 Ci)	=	37 GBq

APPENDIX 2

Correlation of hydrogen chemical shifts and coupling constants with molecular structure

In the following Tables 47 to 53, the chemical shift δ (p.p.m.) values are averages from several compounds and sources, all measured for dilute solutions in deuterochloroform from internal TMS or DSS. Discrepancies from values observed in particular cases might be ± 0.5 p.p.m. but will generally be much less.

The chemical shift is defined by

$$\delta = \frac{(\nu_{\text{sample}} - \nu_{\text{reference}})\ \text{Hz}}{(\nu_{\text{spectrometer}})\ \text{MHz}}\ \text{p.p.m.} \tag{17}$$

For ^{3}H nmr spectroscopy the shifts are conveniently measured from a ghost reference, i.e. the position ($\delta = 0$) equivalent to that of the ^{3}H nmr signal from partially monotritiated TMS (or DSS). This latter position is obtained from the observed internal ^{1}H reference frequency by multiplying by the ratio $\omega_T/\omega_H = 1.06663974$ (see page 7). Such a method of referencing for ^{3}H nmr spectroscopy ensures that ^{3}H chemical shifts are virtually the same as ^{1}H chemical shifts when measured in dilute solution in the usual solvents (preferably $CDCl_3$) (see also page 4).

Table 54 gives correlations between hydrogen–hydrogen spin–spin coupling constants and molecular structure and stereochemistry.

Table 47. Alkyl group shifts (one substituent)

Chemical shifts (δ) for CH_3, CH_2 and CH groups attached directly to a group X and to saturated hydrocarbon residues R, R′, R″:

X	CH_3X	$R'CH_2X$	$R'R''CHX$
–R	0.9	1.25	1.5
$-CH^{b}-CH_2^{a}$ (epoxide, ring O)	1.32	(a) 2.4, 2.7	(b) 3.0
⟍=, ⟩= (R), ⟍= –R	1.7	1.95	2.6
⟍=–=–= etc. (end-of-chain position)	1.82		
=⟍=–= etc. (in-chain position)	1.97	2.2	
–≡–R	1.8		
⟍=N.R.	2.0		
–CO.OR	2.0	2.1	
–CN	2.0	2.48	2.7
$-CO.NH_2$, $-CO.NR_2$	2.02	2.05	
–CO.OH	2.07	2.34	2.57
–CO.R	2.10	2.40	2.48
–SH, –SR	2.10	2.40	3.2
$-NH_2$, $-NR_2$	2.15	2.50	2.87
–I	2.16	3.15	4.20
–CHO	2.17	2.2	2.4
–Ph	2.34	2.62	2.87
–Br	2.65	3.34	4.1
–NH.CO.R	2.9	3.3	3.5
–Cl	3.02	3.44	4.02
–OR	3.30	3.36	3.8
$-\overset{+}{N}R_3$	3.33	3.40	3.50
–OH	3.38	3.56	3.85
$-O.SO_2.OR$	3.58		
–O.CO.R.	3.65	4.15	5.01
–OPh	3.73	3.90	4.00
–O.CO.Ph	3.90	4.23	5.12
$-O.CO.CF_3$	4.10	4.43	
–F	4.26	4.35	
$-NO_2$	4.33	4.40	4.60

Table 48. Alkyl group shifts resulting from an additional β-substituent

If in addition to the X substituent listed in Table 47 there is also a Y substituent β to the CH_3, CH_2 or CH group of interest, then the additional correction is:

Y	β-Shifts		
	CH_3–C–Y	CH_2–C–Y	CH–C–Y
–C=C	+0.1	+0.05	
–CO.OH, –CO.OR	+0.25	+0.4	
–CN	+0.4	+0.4	
–CO.NH_2	+0.23		
–CO.R, –CHO	+0.2	+0.39	
–SH, –SR	+0.45	+0.3	
–NH_2, –NR_2	+0.1	+0.05	
–I	+1.0	+0.5	+0.4
–Ph	+0.35	+0.3	
–Br	+0.8	+0.6	+0.25
–NH.CO.R	+0.1		
–Cl	+0.6	+0.4	+0.2
–OH, –OR	+0.27	+0.1	
–O.CO.R	+0.37		
–O.Ph		+0.35	
–NO_2	+0.67	+0.8	

Table 49. CH_2-group shifts resulting from two substituents

Chemical shifts (δ) for the hydrogens of methylene groups $CH_2 \begin{smallmatrix} \diagup X \\ \diagdown X' \end{smallmatrix}$ (where *neither* substituent is alkyl) are given by $\delta = 1.25 + \Sigma a$, where the a's are the corrections for the X,X′ groups:

X	a	X	a
–=	0.75	–Ph	1.3
–≡	0.9	–Br	1.9
–CO.OH, –CO.OR	0.8	–Cl	2.0
–CN, –CO.R	1.2	–OH, –OR	1.7
–SH, –SR	1.0	–O.CO.R	2.7
–NH_2, –NR_2	1.0	–OPh	2.3
–I	1.4		

Extension to chemical shifts of $CH \begin{smallmatrix} \diagup X \\ -X' \\ \diagdown X'' \end{smallmatrix}$ groups by means of the formula $\delta = 1.50 + \Sigma a$ is less accurate.

Table 50. (a) *Alkenyl hydrogen shifts.*

Chemical shifts (δ) for H attached to unsaturated groups:

H–C≡C–R	1.8	$>\!=\!<$ H, CO$^-$	5.8
H–C≡C(OH)RR′	2.4	H$>\!=\!<$CO$^-$	6.0
H–C≡C–=–	2.8		
H–C≡C–Ph	2.93		
H–C≡C–CO–	3.17	$>\!=\!<$ H, CO$^-$	6.2
$H_2C=CRR'$	4.65		
H_2C=–=–	4.9	Ph–=$<$ H, CO$^-$ (cis or trans)	6.6
HC(R)=C(R′)OR″	5.0		
HC(R)=C=CR′R″ (allenic)	5.2	H, Ph $>$=–CO$^-$	7.8
HC(R)=CR′R″	5.3		
Ph, R $>\!=\!<$ H	5.35	R_2N–CHO	7.85
H	5.05	RO–CHO	8.03
cycloalkene =C–H	5.6	R–CHO	9.65
		Ar–CHO	9.9
–=C(H)–=– (in chain)	6.2		
$>\!=\!<$ H, OR	6.8		
$>\!=\!<$ H, Ph	7.0		

Table 50. (b) *Aryl and heteroaryl hydrogen shifts.*

Benzene: H* 7.27

Naphthalene: H 7.78; H 7.42

Pyridine: H 7.36; H 6.99; H 8.50

Pyridazine: H 7.50; H 9.24

Pyrimidine: H 7.37; H 8.77; H 9.26

Pyrazine: H 8.50

Pyrrole (NH): H 6.06; H 6.53

Thiophene (S): H 7.06; H 7.17

Furan (O): H 6.28; H 7.36

Dihydrofuran (O): H 4.86; H 6.23

Dihydropyran (O): H 4.54; H 6.24

*(For substituent effects see Table 51.)

Table 51. Substituent effects on benzene hydrogen chemical shifts

Shifts in the position of benzene H (δ = 7.27) caused by one or more substituents on the ring (negative shifts are to higher field; positive shifts are to lower field):

Substituent	Ortho	Meta	Para
$-CH_3$	−0.15	−0.1	−0.1
−=	+0.2	+0.2	+0.2
−CHO	+0.65	+0.2	+0.3
−CO.R	+0.6	+0.3	+0.3
−CO.OH, −CO.OR	+0.8	+0.15	+0.2
$-CO.NH_2$	+0.5	+0.2	+0.2
−C≡N	+0.3	+0.3	+0.3
−Cl	0	0	0
−Br	+0.2	−0.2	0
−I	+0.3	−0.2	−0.1
$-NH_2$	−0.8	−0.15	−0.4
$-N(CH_3)_2$	−0.5	−0.2	−0.5
$-\overset{+}{N}H_3$	+0.4	+0.2	+0.2
−NH.CO.R	+0.4	−0.2	−0.3
$-NO_2$	+1.0	+0.3	+0.4
−OH	−0.4	−0.1	−0.4
−OR	−0.4	−0.1	−0.4
−O.CO.R	+0.2	−0.1	−0.2
−SR	+0.1	−0.1	−0.2
$-SO_2.NH_2$	+0.55	+0.15	+0.3

Table 52. Chemical shift ranges for exchangeable hydrogens

These signals may be broadened, and the positions are dependent upon concentration, solvent, temperature and the degree of hydrogen bonding:

Compound	δ	Trends
R–OH	0.5 to 4.0	H-bonded enols, 11 to 18
Ph–OH	4.5 to 9.0 or so	Shift increased by H-bonding
R–CO.OH	8 to 13 or so	
$R-NH_2$	1 to 5	} Shift increased by H-bonding
$Ph-NH_2$	3.5 to 6 or more	}
$R-CO.NH_2$	5 to 8.5	Often rather broad
R–CO.NH.CO–R	9 to 12	
R–SH	1 to 3	} Shift increased by H-bonding
Ph–SH	3.5 or more	}
=N–OH	10 to 12 or so	May be broad

Table 53. Solvent resonances

Chemical shifts for protic solvents (line positions vary a little with solute):

Solvent	δ	Solvent	δ
$CF_3.CO_2H$	9.83	Dioxan	3.68
$H.CO.NMe_2$	7.85	$CH_3.OH$	3.47 (CH_3)
$CHCl_3$	7.27	$[CH_3]_2SO$	2.60
C_6H_6	7.27	$[CH_3]_2CO$	2.17
$CHBrCl_2$	7.20	$CH_3.CO_2H$	2.1 (CH_3)
$CHBr_3$	6.85		11.6 (OH)
CH_2Cl_2	5.30	$CH_3.CN$	2.00
H_2O (or HOD)	*ca.* 4.8	Cyclohexane	1.43
$CH_3.NO_2$	4.33	$[CH_3]_3COH$	1.27 (CH_3)

The intense solvent signal is accompanied by spinning side bands so that a region up to ± 1 p.p.m. on each side of the solvent resonance is obscured.

Table 54. Coupling constants

Selection of proton–proton spin-coupling constants: corresponding triton–proton and triton–triton spin-coupling constants are estimated from the expressions $J(\mathrm{T,H}) = J(\mathrm{H,H}) \times 1.06664$ and $J(\mathrm{T,T}) = J(\mathrm{H,H}) \times (1.06664)^2$ (see page 9).

Function	Type of coupling	J_{AB} (Hz)	Reference
Methanes			
$>C<^{HA}_{HB}$	Geminal	-10 to -18	
Ethanes			
CH_3–CH_2X (A B)	Vicinal (free rotation)	5 to 10 $J_{Average} = 8.4 - 0.4E$ where E is the Huggins electronegativity X	[208]
CH_3–C(=O)H (A: CH_3, B: H)		2.9	
$>CH$–$CH<$ (A B)	Vicinal (no rotation)	0 to 12 depending on dihedral angle	
Vinyl derivatives			
$>C=C<^{H_A}_{H_B}$	Geminal	-3 to $+7$	
$\diagdown N=<^{H_A}_{H_B}$	Geminal	-8 to -16	
Ethenes			
$>C=C<$ (H_A, H_B)	Cis	Acyclic or 6-ring: 5 to 14 In 5-ring: 3 to 6 In 7-ring: 9 to 14	

Table 54 (*contd.*)

Function	Type of coupling	J_{AB} (Hz)				Reference
H_A >C=C< H_B	Trans	11 to 19				
>C=CH_A–CH_B=C<	Vicinal	10 to 13				
Allylic –CH_A=C–CH_B	Cis or trans	−2 to +1				
>C=C<CH_A H_B		4 to 10				
	Miscellaneous long range:					
H_A–C–C=C–C–H_B	Homoallylic	0 to 2				
H_AC–C≡C–C–H_B		2 to 3				
H_A–C–C≡C–H_B		−2 to −3				
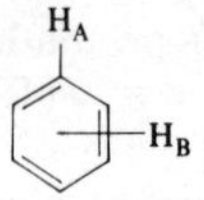 H_A, H_B (benzene ring)	Benzenoid	Ortho—7 to 10 Meta—2 to 3 Para—0 to 1				
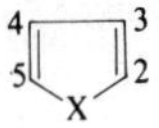 5-membered ring (X; positions 2, 3, 4, 5)	Heterocyclic	$J_{2,3}$	$J_{3,4}$	$J_{2,4}$	$J_{2,5}$	
	X = O, furan	1.8	3.4	0.8	1.6	
	= S, thiophen	5.2	3.6	1.3	2.7	
	= NH, pyrrole	2.6	3.4	1.4	2.1	
Pyridine ring (N; positions 2, 3, 4, 5, 6)	Pyridine	5.5	7.5	1.9	0.9	

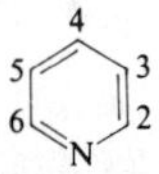

References and bibliography

[1] E. A. Evans, *Tritium and Its Compounds*, 2nd ed., Butterworths, London, 1974.

[1a] H. L. Anderson and A. Novick, Magnetic moment of the triton, *Phys. Rev.*, **71**, 372 (1947); F. Bloch, A. C. Graves, M. Packard and R. W. Spence, Spin and magnetic moment of tritium, *Phys. Rev.*, **71**, 373 (1947).

[1b] F. Bloch, A. C. Graves, M. Packard and R. W. Spence, Relative moments of H_1 and H_3, *Phys. Rev.*, **71**, 551 (1947); R. W. Huggins and J. H. Sanders, Nuclear magnetic moment ratios measured in high and low fields, *Proc. Phys. Soc.*, **86**, 53 (1965).

[2] P. J. Ayres and E. A. Evans, Unpublished experiments, 1964, Middlesex Hospital Medical School and Amersham (The Radiochemical Centre).

[3] G. V. D. Tiers, C. A. Brown, R. A. Jackson and T. N. Lahr, Tritium nuclear magnetic resonance spectroscopy 1. Observation of high-resolution signals from the methyl and methylene groups of ethylbenzene. The non-radiochemical use of tritium as a tracer, *J. Am. Chem. Soc.*, **86**, 2526 (1964).

[4] J. Bloxsidge, J. A. Elvidge, J. R. Jones and E. A. Evans, Tritium nuclear magnetic resonance spectroscopy. Part 1: Technique, internal referencing and some preliminary results, *Org. Mag. Res.*, **3**, 127 (1971).

[5] J. M. A. Al-Rawi, J. P. Bloxsidge, C. O'Brien, D. E. Caddy, J. A. Elvidge, J. R. Jones and E. A. Evans, Tritium nuclear magnetic resonance spectroscopy. Part 2. Chemical shifts, referencing and an application, *J.C.S. Perkin II*, 1635 (1974).

[6] R. K. Harris, Nuclear magnetic resonance and the periodic table, *Chem. Soc. Rev.*, **5**, 1 (1976).

[7] R. C. Weast (Ed.), *Handbook of Chemistry and Physics*, 62nd ed., Chemical Rubber Co. Press, Cleveland, 1981–1982.

[8] C. B. Taylor and W. Roether, A uniform scale for reporting low-level tritium measurements in water, *Intl. J. Appl. Radiat. Isotopes*. **33,** 377 (1982); IAEA-TECDOC-246 Proc. Consultants meeting on low level tritium measurement, IAEA, Vienna, 24–28 September, 1979.

[9] W. Duffy, Magnetic moment of the triton in units of the magnetic moment of the proton, *Phys. Rev.*, **115,** 1012 (1959).

[10] P. Diehl and T. Leipert, Deuteronen-Kernresonanzspektroskopie, *Helv. Chim. Acta*, **47,** 545 (1964).

[11] R. Price, Ph.D. Thesis, University of London, 1969.

[12] E. W. Randell and D. G. Gillies, Nitrogen nuclear magnetic resonance, *Progr. in Nmr Spectroscopy*, **3,** 119 (1971).

[13] University of Surrey, Unpublished results.

[14] J. P. Bloxsidge, J. A. Elvidge, J. R. Jones, R. B. Mane and M. Saljoughian, Tritium nuclear magnetic resonance spectroscopy. Part 11. Further consideration of referencing, isotope effects and coupling constants: Preparation of [^{3}H]tetramethylsilane, *Org. Mag. Res.*, **12,** 574 (1979).

[15] W. Saur, H. L. Crespi and J. J. Katz, Vicinal deuterium isotope effects on proton chemical shifts, *J. Mag. Res.*, **2,** 47 (1970).

[16] A. L. Allred and W. D. Wilk, Long-range deuterium isotope effects on chemical shifts, *J. C. S. Chem. Comm.*, 273 (1969).

[17] R. M. Lynden-Bell and R. K. Harris, *Nuclear Magnetic Resonance Spectroscopy*, Nelson, London, 1969.

[18] J. M. A. Al-Rawi, J. A. Elvidge, J. R. Jones and E. A. Evans, Tritium nuclear magnetic resonance spectroscopy. Part 3. Coupling constants and isotope effects, and calculation of $^2J_{HH}$ coupling constants, *J.C.S. Perkin II*, 449 (1975).

[19] J. P. Bloxsidge, J. A. Elvidge, J. R. Jones, E. A. Evans, J. P. Kitcher and D. C. Warrell, Tritium nuclear magnetic resonance spectroscopy. Part 14. Analysis of tritiated methyl groups, *Org. Mag. Res.*, **15,** 214 (1981).

[20] J. M. A. Al-Rawi, J. P. Bloxsidge, J. A. Elvidge, J. R. Jones, V. E. M. Chambers, V. M. A. Chambers and E. A. Evans, Tritium nuclear magnetic resonance spectroscopy. Part 6. Tritiated steroid hormones, *Steroids*, **28,** 359 (1976).

[21] Y. Osawa and D. G. Spaeth, Estrogen biosynthesis. 1. Estrogen biosynthesis. Stereospecific distribution of tritium in testosterone-1α,2α-t_2, *Biochemistry*, **10,** 66 (1971).

[22] J. Fishman, H. Guzik and D. Dixon, Stereochemistry of Estrogen Biosynthesis, *Biochemistry*, **8,** 4304 (1969).

[23] E. A. Evans, J. A. Elvidge, J. R. Jones and D. C. Warrell, Unpublished results.

[24] P. A. Bell and E. Kodicek, The stereospecificity of tritium distribution in [1-^{3}H]- and [1,2-$^3H_2$]-cholesterol and -cholecalciferol, *Biochem. J.*, **116,** 755 (1970).

[25] J. M. A. Al-Rawi, J. A. Elvidge, J. R. Jones, V. M. A. Chambers and E. A. Evans, Tritium nuclear magnetic resonance spectroscopy. Part 4. Distribution of tritium in [G-^{3}H]phenylalanine and other amino acids, *J. Label. Compounds and Radiopharmaceuticals*, **12,** 265 (1976).

[26] M. M. Clifford, E. A. Evans, A. E. Kilner and D. C. Warrell, Distribution of tritium label in DL-[G-^{3}H]phenylalanine, *J. Label. Compounds*, **11,** 435 (1975).

[27] J. M. A. Al-Rawi, J. P. Bloxsidge, J. A. Elvidge, J. R. Jones, V. M. A. Chambers and E. A. Evans, Tritium nuclear magnetic resonance spectroscopy. Part 5. Distribution of tritium in labelled polycyclic hydrocarbons, *J. Label. Compounds and Radiopharmaceuticals*, **12,** 293 (1976).

[28] W. P. Duncan and J. F. Engel, Labeled metabolites of polycyclic aromatic hydrocarbons. 1. Determination of tritium at the 6-position in benzo[a]pyrene-G-^{3}H, *J. Label. Compounds*, **11,** 145 (1975).

[29] J. P. Bloxsidge, J. A. Elvidge, J. R. Jones, R. B. Mane and E. A. Evans, Tritium nuclear magnetic resonance spectroscopy, Part 8. Significance of the nuclear Overhauser effect for quantitative analysis of tritium, *J. Chem. Research* (*S*), 258 (1977).

[30] B. W. Bycroft, University of Nottingham, personal communication.

[31] J. H. Noggle and R. E. Schirmer, *The Nuclear Overhauser Effect: Chemical Applications*, Academic Press, New York, 1971.

[32] L. J. Altman and N. Silberman, Tritium nuclear magnetic resonance spectroscopy. Distribution patterns and nuclear Overhauser enhancements in some tritiated steroids, *Steroids*, **29,** 557 (1977).

[33] W. McFarlane and D. S. Rycroft, Multiple resonance, Chapter 10 in *Nuclear Magnetic Resonance*, Royal Society of Chemistry, London, 1981.

[34] D. Calvert, R. W. Martin and A. L. Odell, Chemical shifts for tritons in the ortho, meta and para positions of toluene as determined by tritium nmr spectroscopy, *Org. Mag. Res.*, **11,** 213 (1978).

[35] M. A. Long, J. L. Garnett and J. C. West, ^{3}H N.M.R. studies of aromatics by boron tribromide and aluminium chloride catalysed tritiated water exchange, *Tetrahedron Lett.*, 4171 (1978).

[36] L. J. Altman and L. Thomas, Determination of the specific activity of high specific activity tritium labeled sodium borohydride by tritium nuclear magnetic resonance spectroscopy, *Anal. Chem.*, **52,** 992 (1980).

[37] D. J. Aberhart and C-H Tann, Application of tritium nmr spectroscopy in the determination of the stereochemistry of dehydrogenation of isobutyryl CoA in *Pseudomonas putida*, *J. Am. Chem. Soc.*, **102,** 6377 (1980).

[38] H. Levine-Pinto, J. L. Morgat, P. Fromageot, D. Meyer, J. C. Poulin and H. B. Kagan, Asymmetric tritiation of *N*-acetyl dehydrophenylalanyl-(*S*)phenylalanine (methyl ester) catalysed with a rhodium (+)diop complex, *Tetrahedron*, **38,** 119 (1980).

[39] D. H. G. Crout, M. Lutstorf, P. J. Morgan, R. M. Adlington, J. E. Baldwin and M. J. Crimmin, Unusual stereospecificity in the hydrogenation of an isopropenyl function with Wilkinson's catalyst; A route to chiral methyl valine, *J. C. S. Chem. Comm.*, 1175 (1981).

[40] J. A. Elvidge, Tritium nmr spectroscopy and applications, pp. 35–44 in *Proc. Intl. Symp. on Synthesis and Applications of isotopically Labeled Compounds, Kansas City, 6–11 June 1982* (Eds: W. P. Duncan and A. R. Susan), Elsevier, Amsterdam, 1983.

[41] R. R. Ernst, Sensitivity enhancement in magnetic resonance, *Adv. Mag. Res.*, **2,** 1 (1966).

[42] J. P. Bloxsidge and J. A. Elvidge, Practical aspects of tritium magnetic resonance, *Prog. NMR Spectroscopy*, **16,** 99 (1983).

[43] R. A. Flath, N. Henderson, R. E. Lundin and R. Teranishi, Microcell for nuclear magnetic resonance analysis, *Appl. Spectroscopy*, **21,** 183 (1969).

[44] J. M. A. Al-Rawi and J. P. Bloxsidge, A general purpose n.m.r. microcell system suitable for volatile, air sensitive and toxic samples, *Org. Mag. Res.*, **10,** 261 (1977).

[45] S. P. Sawan and T. L. James, Use of a plastic tube insert for nmr experiments entailing hazardous samples, *J. Mag. Res.*, **32,** 173 (1978).

[46] M. Muramatsu, Y. Suzuki, I. Miyanaga and Y. Wadachi, Safety aspects of radioactive experiments, Chapter 3 in *Radiotracer Techniques and Applications* (Eds. E. A. Evans and M. Muramatsu), Vol. 1, M. Dekker Inc., New York, 1977.

[47] W. D. Chiswell and G. H. C. Dancer, Measurement of tritium concentration in exhaled water vapor as a means of estimating body burdens, *Health Phys.*, **17,** 331 (1969).

[48] cf. D. G. Ott, V. N. Kerr, T. W. Whaley, T. Benzinger and R. K. Rohmer, Syntheses with stable isotopes: Methanol-^{13}C, methanol-^{13}C-^{2}H$_4$ and methanol-^{12}C, *J. Label. Compounds*, **10,** 315 (1974).

[49] J. P. Bloxsidge, J. A. Elvidge, M. Gower, J. R. Jones, E. A. Evans, J. P. Kitcher and D. C. Warrell, Tritium nuclear magnetic resonance spectroscopy. Part 13. Tritium labelled neurochemicals, *J. Label. Compounds and Radiopharmaceuticals*, **18,** 1141 (1981).

[50] P. A. Bell and W. P. Scott, A synthesis of 25-hydroxycholecalciferol-(26,27-^{3}H), *J. Label. Compounds*, **9,** 339 (1973).

[51] R. F. Glascock and L. R. Reinius, Studies on the origin of milk fat. 1. The location of tritium in stearic acid produced by the catalytic addition of tritium to elaidic acid, *Biochem. J.*, **62,** 529 (1956).

[52] S. F. Sakrzewski, E. A. Evans and R. F. Phillips, On the specificity of labeling in tritiated folic acid, *Analyt. Biochem.*, **36,** 197 (1970).

[53] E. A. Evans, J. P. Kitcher, D. C. Warrell, J. A. Elvidge, J. R. Jones and R. M. Lenk, Tritium nuclear magnetic resonance spectroscopy. Part 12. Patterns of labelling in tritiated folic acid and methotrexate, *J. Label. Compounds and Radiopharmaceuticals*, **16,** 697 (1979).

[54] J. A. Elvidge, Tritium nuclear magnetic resonance spectroscopy, Chapter 9 in *The Multinuclear Approach to NMR Spectroscopy* (Eds.) J. B. Lambert and F. G. Riddell, D. Reidel, London, 1983.

[55] R. Voges, H. Andres, H. R. Loosli and E. Schreier, Use of ^{3}H-nmr spectroscopy as an aid in the evaluation process for the most advantageous tritium labelling procedure, pp. 331–336 in *Proc. Intl. Symp. on Synthesis and Applications of*

Isotopically Labeled Compounds, Kansas City, 6–11 June 1982 (Eds: W. P. Duncan and A. B. Susan), Elsevier, Amsterdam, 1983.

[56] K. E. Wilzbach, Tritium labeling by exposure of organic compounds to tritium gas, *J. Am. Chem. Soc.*, **79,** 1013 (1957).

[57] E. A. Evans, H. C. Sheppard, J. C. Turner and D. C. Warrell, A new approach to specific labelling of organic compounds with tritium: Catalysed exchange in solution with tritium gas, *J. Label. Compounds*, **10,** 569 (1974).

[58] O. Buchman, I. Pri-Bar and M. Shimoni, Catalysed exchange between tritium gas and organic molecules in solution, *J. Label. Compounds and Radiopharmaceutical*, **14,** 155 (1978).

[59] M. Kaspersen, F. M. van Rooy, J. Wallaart and C. Funke, *Rec. Trav. Chim., Pays-Bas*, **102,** 450 (1983).

[60] I. Pri-Bar and O. Buchman, Hydrogen exchange between molecular tritium and bibenzyl in solution, catalyzed by transition metals, *Int. J. Appl. Rad. Isotopes*, **27,** 53 (1976).

[61] J. R. Jones, Labelling of molecules, Chapter 11 in *The Ionisation of Carbon Acids*, Academic Press, London, 1973.

[62] J. P. Bloxsidge, J. A. Elvidge, J. R. Jones, R. B. Mane, V. M. A. Chambers, E. A. Evans and D. Greenslade, [G-^{3}H]Vinblastine and its analysis by tritium nuclear magnetic resonance spectroscopy, *J. Chem. Research* (*S*), 42 (1977).

[63] R. O. C. Norman and R. Taylor, *Electrophilic Substitution in Benzenoid Compounds*, Elsevier, Amsterdam, 1965.

[64] M. Orchin and D. M. Bollinger, Hydrogen–deuterium exchange in aromatic compounds, *Structure and Bonding*, **23,** 167 (1975).

[65] P. C. Crossley, R. W. Martin, J. B. Mawson and A. L. Odell, Tritium labelling and tritium nmr. Part 1. Alpha-labelled stearic, palmitic, myristic and lauric acids, *J. Label. Compounds and Radiopharmaceuticals*, **17,** 779 (1980).

[66] N. H. Werstiuk and T. Kadai, Acid-catalysed deuterium exchange, pp. 13–20 in *Proc. 1st Intl. Conf. on Stable Isotopes in Chemistry, Biology and Medicine* (Eds. P. D. Klein and S. V. Peterson), Oak Ridge, Tennessee, 9–11 May 1973; USAEC. Conf-730525.

[67] M. A. Long, J. L. Garnett and R. F. W. Vining, Rapid deuteriation and tritiation of organic compounds using organometallic and elemental halides as catalysts, *J. C. S. Perkin II*, 1298 (1975).

[68] J. A. Elvidge, J. R. Jones, M. A. Long and R. B. Mane, ^{3}H nmr. Demonstration of the pattern of tritiation of alkanes by ethylaluminium dichloride/tritiated water, *Tetrahedron Lett.*, 4349 (1977).

[69] E. A. Evans, R. H. Green, J. A. Spanner and W. R. Waterfield, Labilization of the α-hydrogen atom of generally labelled tritiated L-α-amino acids in the presence of renal D-amino acid oxidase, *Nature*, **198,** 1301 (1963).

[70] L. J. Altman and N. Silberman, ^{3}H nmr analysis of labeled prolines, *Analyt. Biochem.*, **79,** 302 (1977).

[71] R. B. Herbert and I. T. Nicolson, Commercial [G-^{3}H]-DL-phenylalanine: Distribution of label, *J. Label. Compounds*, **9,** 567 (1973).

[72] J. W. Emsley, L. Phillips and V. Wray, Fluorine coupling constants, *Prog. NMR Spectroscopy*, **10,** 83 (1977).

[73] H. Simon and O. Berngruber, Mechanistic studies on the catalytic hydrogenation of α,β-unsaturated carbonyl compounds with platinum or palladium, based on intramolecular tritium distribution, *Tetrahedron Lett.*, 4711 (1968).

[74] E. A. Evans, D. C. Warrell, J. A. Elvidge and J. R. Jones, Catalytic tritiation studies using tritium nmr spectroscopy, *J. Radioanalyt. Chem.*, **64,** 41 (1981).

[75] D. J. Aberhart and H-J. Lin, Syntheses of (3*RS*)-[2,3-$^3H_2$(*N*)]-β-leucine, *J. Label. Compounds and Radiopharmaceuticals*, **20,** 611 (1983).

[76] J. A. Elvidge, J. R. Jones, M. S. Saieed, E. A. Evans and D. C. Warrell, Mechanistic aspects of the tritiation and deuteriation of taurine, *J. Chem. Research* (*S*), 288 (1981).

[77] H. Levine-Pinto, J. L. Morgat, P. Fromageot, D. Meyer, J. C. Poulin and H. B. Kagan, Asymmetric tritiation of *N*-acetyldehydrophenylalanyl(*S*)phenylalanine (methyl ester) catalysed with a rhodium (+)diop complex, *Tetrahedron*, **38**, 119 (1982).

[78] J. A. Elvidge, J. R. Jones, R. B. Mane, V. M. A. Chambers, E. A. Evans and D. C. Warrell, Tritium nuclear magnetic resonance spectroscopy. Part 9. Specifically tritiated glucoses, *J. Label. Compounds and Radiopharmaceuticals*, **15**, 141 (1978).

[79] K. Schmidt, J. Genovese and J. Katz, Enzymatic degradation of isotopically labelled compounds. 11. Glucose labeled with ^{14}C and tritium, *Analyt. Biochem.*, **34**, 170 (1970).

[80] J. Katz, P. A. Wals, S. Golden and R. Rognstad, Recycling of glucose by rat hepatocytes, *Eur. J. Biochem.*, **60**, 91 (1975).

[81] J. E. G. Barnett and D. L. Corina, A synthesis of D-glucose-5-*t* and D-glucose-5-*t*-6-cf. phosphate, *Carbohydrate Res.*, **3**, 134 (1966).

[82] P. M. Collins, The kinetics of the acid catalysed hydrolysis of some isopropylidene furanoses, *Tetrahedron*, **21**, 1809 (1965).

[83] O. Gabriel, A facile synthesis of tritiated D-glucose and D-galactose labeled uniquely at carbon 4, *Carbohydrate Res.*, **6**, 319 (1962).

[84] W. Mackie and A. S. Perlin, 1,2-*O*-isopropylidene-α-glucofuranose-5-*d* and -5,6,6′-d_3, *Canad. J. Chem.*, **43**, 2921 (1965).

[85] H. S. Isbell, H. L. Frush and J. D. Moyer, Tritium-labeled compounds. 4. D-Glucose-6-*t*, D-xylose-5-*t*, and D-mannitol-1-*t*, *J. Res. Natl. Bur. Stds.*, **64a**, 359 (1960).

[86] A. De Bruyn and M. Anteunis, ^{1}H NMR study of 2-deoxy-D-*Arabino*-hexopyranose (2-deoxy glucopyranose), 2-deoxy-D-*Lyxo*-hexopyranose (2-deoxy galactopyranose) and 2′-deoxylactose. Shift increment studies in 2-deoxy carbohydrates, *Bull. Soc. Chim. Belges*, **84**, 1201 (1975).

[87] J. A. Elvidge, J. R. Jones, R. M. Lenk, Y. S. Tang, E. A. Evans, G. L. Guilford and D. C. Warrell, Mechanistic aspects of hydrogenation reactions as studied by ^{3}H nuclear magnetic resonance spectroscopy, *J. Chem. Research* (*S*), 82 (1982).

[88] C. W. Haigh and R. B. Mallion, Proton magnetic resonance of planar condensed benzenoid hydrocarbons. 1. Analysis of spectra, *Mol. Phys.*, **18**, 737 (1970).

[89] K. D. Bartle, D. W. Jones and R. S. Mathews, High-field nuclear magnetic resonance spectra of some carcinogenic polynuclear hydrocarbons, *Spectrochim. Acta.*, **25a**, 1603 (1969).

[90] C. W. Haigh and R. B. Mallion, Proton magnetic resonance of 3,4-benzopyrene at 100 and 220 Mcps, *J. Mol. Spectroscopy*, **29**, 478 (1969).

[91] We thank Dr. O. Howarth (University of Warwick) for these spectra.

[92] W. P. Duncan, Communication. The compound was prepared at the Midwest Research Institute as part of the Cancer Research Program of the National Cancer Institute, Bethesda, Maryland, USA.

[93] J. M. A. Al-Rawi, J. A. Elvidge, J. R. Jones, R. B. Mane and M. Saieed, Selectivity of catalysts for hydrogen isotope exchange as studied by ^{3}H nmr spectroscopy, *J. Chem. Research* (*S*), 298 (1980).

[94] M. A. Long, J. L. Garnett, P. G. Williams and T. Mole, Tritium labeling of organic compounds by HNaY zeolite catalyzed exchange with tritiated water and their analysis by ^{3}H nmr, *J. Am. Chem. Soc.*, **103**, 1571 (1981).

[95] M. A. Long and C. A. Lukey, Chemical shifts for tritons in chlorobenzene, bromobenzene, fluorobenzene and toluene as determined by tritium nmr spectroscopy, *Org. Mag. Res.*, **12**, 440 (1979).

[96] G. L. Guilford, E. A. Evans, D. C. Warrell, J. R. Jones, J. A. Elvidge, R. M. Lenk and Y. S. Tang, Studies of catalysis, pp. 327–330 in *Proc. Intl Symp. on Synthesis and Applications of Isotopically Labeled Compounds, Kansas City, 6–11 June 1982* (Eds. W. P. Duncan and A. B. Susan), Elsevier, Amsterdam, 1983.

[97] M. A. Long, J. L. Garnett and P. G. Williams, Tritium nmr spectroscopy of compounds labelled by exchange over zeolite and metal catalysts, pp. 315—320 in *Proc. Intl. Symp. on Synthesis and Applications of Isotopically Labelled Compounds, Kansas City*, 6—11 *June*, 1982 (Eds. W. P. Duncan and A. B. Susan), Elsevier, Amsterdam, 1983.

[98] B. Clin, J de Bony, P. Lalanne, J. Bais and B. Lemanceau, A new scheme for resolution enhancement in Fourier transform nmr, *J. Mag. Res.*, **33,** 457 (1979).

[98a] C. D. Jardetzky and O. Jardetzky, Investigation of the Structure of Purines, Pyrimidines, Ribose Nucleosides and Nucleotides by Proton Magnetic Resonance. II, *J. Am. Chem. Soc.*, **82,** 222 (1960); S. S. Danyluk and F. E. Hruska, The Effect of pH upon the Nuclear Magnetic Resonance Spectra of Nucleosides and Nucleotides, *Biochemistry*, **7,** 1038 (1968); O. Takaku, H. Haraguchi, S. Toda and K. Fuwa, A Study on 5′-Nucleotides in D_2O Solution by Proton Magnetic Resonance Spectroscopy, *Agr. Biol. Chem.*, **39,** 2373 (1975); M. P. Schweizer, A. D. Broom, P. O. P. T'so and D. P. Hollis, Studies of Inter- and Intra-molecular Interaction in Mononucleotides by Proton Magnetic Resonance, *J. Am. Chem. Soc.*, **90,** 1042 (1968).

[99] A. De Marco and K. Wüthrich, Digital filtering with a sinusoidal window function: An alternative technique for resolution enhancement in FT nmr, *J. Mag. Res.*, **24,** 207 (1976).

[100] S. E. Taylor, Ph.D. Thesis, University of Surrey, 1978.

[101] H. J. Brodie, K. Raab, G. Possanza, N. Seto and M. Gut, Further stereochemical studies of the catalytic reduction of $\Delta^{1,4}$-3-keto steroids with tritium, *J. Org. Chem.*, **34,** 2697 (1969).

[102] J. Fishman, H. Guzik and D. Dixon, Stereochemistry of estrogen biosynthesis, *Biochemistry*, **8,** 4304 (1969).

[103] J. M. A. Al-Rawi, J. A. Elvidge, R. Thomas and B. J. Wright, Direct determination of the stereospecificity of the Δ^1-dehydrogenation of testosterone using tritium nuclear magnetic resonance, *J.C.S. Chem. Comm.*, 1031 (1974).

[104] J. A. Elvidge, Deuterium and tritium nuclear magnetic resonance spectroscopy, Chapter 5 in *Isotopes: Essential Chemistry and Applications* (Eds. J. A. Elvidge and J. R. Jones), The Chemical Society, London, 1980.

[105] L. J. Altman, C. Y. Han, A. Bertolino, G. Handy, D. Laungani, W. Muller, S. Schwartz, D. Shanker, W. H. de Wolf and F. Yang, Stereochemistry of the 1,3-proton loss from a chiral methyl group in the biosynthesis of cyclo-artenol as determined by tritium nuclear magnetic resonance spectroscopy, *J. Am. Chem. Soc.*, **100,** 3235 (1978).

[106] C. W. Funke, F. M. Kaspersen, J. Wallaart and S. N. Wagenaars, Tritium nmr spectroscopy of steroids, *J. Label Compounds and Radiopharmaceuticals*, **20,** 843 (1983).

[107] J. A. Elvidge, J. R. Jones and M. Saljoughian, Catalytic tritiation of drugs and analysis of the tritium distribution by ^{3}H n.m.r spectroscopy, *J. Pharm. Pharmacol.*, **31,** 508 (1979).

[108] J. A. Elvidge, J. R. Jones, R. B. Mane and J. M. A. Al-Rawi, Tritium nuclear magnetic resonance spectroscopy. Part 10. Distribution of tritium in some labelled nitrogen heterocyclic compounds, *J.C.S. Perkin II*, 386 (1979).

[109] J. A. Elvidge, J. R. Jones, R. B. Mane and M. Saljoughian, Polyene acids. Part 11. Preparation of α-tritiated (or deuteriated) conjugated enoic and dienoic acids and their examination by triton magnetic resonance spectroscopy, *J.C.S. Perkin I*, 1191 (1978).

[110] M. M. Coombs, Tritium labelling of carcinogenic cyclopenta[a]phenanthrenes at

very high specific activity, *J. Label. Compounds and Radiopharmaceuticals*, **17,** 147 (1980).

[111] E. Buncel, J. A. Elvidge, J. R. Jones and K. T. Walkin, Tritium exchange and nuclear magnetic resonance studies on 1,3-dinitrobenzene, *J. Chem. Res.* (*S*), 272 (1980).

[112] E. Buncel, A. R. Norris, J. A. Elvidge, J. R. Jones and K. T. Walkin, Tritium exchange and nuclear magnetic resonance studies on 1,3-dinitronaphthalene, *J. Chem. Res* (*S*), 326 (1980).

[113] C. J. O'Connor, R. W. Martin and D. J. Calvert, Substituent effects on the nmr spectra of carboxylic acid derivatives 11. ^{3}H nmr spectra of substituted benzamides and *N*-alkylbenzamides, *Aust. J. Chem.*, **34,** 2297 (1981).

[114] J. S. Favier, J. Wallaart and F. M. Kaspersen, Tritiation of mianserin, *J. Label. Compounds and Radiopharmaceuticals*, **19,** 1125 (1982).

[115] J. A. Elvidge, D. K. Jaiswal, J. R. Jones and R. Thomas, Tritium nuclear magnetic resonance spectroscopy. Part 7. New information from the tritium distribution in biosynthetically labelled penicillic acid, *J.C.S. Perkin I*, 1080 (1977).

[116] J. M. A. Al-Rawi, J. A. Elvidge, D. K. Jaiswal, J. R. Jones and R. Thomas, Use of tritium nuclear magnetic resonance for the direct location of ^{3}H in biosynthetically-labelled penicillic acid, *J.C.S. Chem. Comm.*, 220 (1974).

[116a] G. Guillaumet and G. Coudert, Synthese d'un antogoniste alpha adrenergique marque au tritium: WB 4101 [benzodioxanyl-1,4-^{3}H-2,3] or *N*-[dimethoxy-2,6-phenoxyethyl]aminomethyl-2 benzodioxanne-1,4[^{3}H-2,3], *J. Labelled Compounds and Radiopharmaceuticals*, **21,** 161 (1984).

[117] G. Pepeu, M. J. Kuhar and S. T. Enna (Eds.), Receptors for neurotransmitters and peptide hormones, *Adv. Biochem. Psychopharm.*, **21,** 1 (1980).

[118] D. J. Aberhart, Application of tritium nmr spectroscopy in the determination of enzyme stereochemistry, pp. 309–314 in *Proc. Intl. Symp. on Synthesis and Applications of Isotopically Labeled Compounds, Kansas City, 6–11 June 1982* (Eds. W. P. Duncan and R. B. Susan), Elsevier, Amsterdam, 1983.

[119] A. L. Odell, K. T. Ronaldson, R. W. Martin and D. J. Calvert, Tritium labelling and tritium nmr. Part 2. Labelled methyl (*Z*)-octadec-9-enoate obtained by partial hydrogenation on lindlar catalyst and its reduction to (*Z*)-octadec-9-en-1-ol, *Aust. J. Chem.*, **35,** 1615 (1982).

[120] P-L. Chien, M-S. Sung and D. B. Bailey, Synthesis of 13-*cis*-[11-^{3}H]-retinoic acid, *J. Label. Compounds and Radiopharmaceuticals*, **16,** 791 (1979).

[121] G. Guroff, J. W. Daly, D. M. Jerina, J. Renson, B. Witkop and S. Udenfriend, Hydroxylation-induced migration: The NIH shift, *Science*, **157,** 1524 (1967).

[122] D. Warshawsky and M. Calvin, Tritium incorporation at specific positions in benzo[a]pyrene, *Biochem. Biophys. Res. Comm.*, **63,** 541 (1975).

[123] G. M. Blackburn, P. E. Taussig and J. P. Will, Tritium incorporation at specific positions in benzo[a]pyrene, *J.C.S. Chem. Comm.*, 907 (1974).

[124] G. M. Blackburn, L. Orgee and G. M. Williams, Oxidative binding of natural oestrogens to DNA by chemical and metabolic means, *J.C.S. Chem. Comm.*, 386 (1977).

[125] K. T. Mason, G. J. Shaw and E. Katz, Biosynthetic studies with L-[2,3-$^3H_2$]valine as precursor of the D-valine moiety in actinomycin, *Arch. Biochem. Biophys.*, **180,** 509 (1977).

[126] F. H. Hulcher, W. H. Oleson and H. B. Lofland, Cholesterol-7-hydroxylase of pigeon liver microsomes, *Arch. Biochem. Biophys.*, **165,** 313 (1974).

[127] R. L. E. Ehrenkaufer, A. P. Wolf and W. C. Hembree, A novel surface for high specific activity tritium labelling using microwave discharge activation of tritium gas, *J. Label. Compounds and Radiopharmaceuticals*, **14,** 271 (1978).

[128] M. H. Randall, L. J. Altman and R. J. Lefkowitz, Structure and biological activity of (-)[^{3}H]dihydroalprenolol—A radioligand for studies of *β*-adrenergic receptors, *J. Med. Chem.*, **20,** 1090 (1977).

[129] For leading references see M. J. Garson and J. Staunton, N.m.r methods for tracing

the fate of hydrogen in biosynthesis, *Chem. Soc. Revs.*, **8,** 539 (1979).

[130] H. J. Ringold, M. Hayano and V. Stefanovic, Concerning the stereochemistry and mechanism of the bacterial C-1,2 dehydrogenation of steroids, *J. Biol. Chem.*, **238,** 1960 (1963).

[131] H. G. Floss and M. D. Tsai, Chiral methyl groups, *Adv. Enzymology*, **50,** 243 (1979).

[132] E. Abraham, C-P. Pang, R. L. White, D. H. G. Crout, M. Lutstorf, P. J. Morgan and A. E. Derome, The stereochemistry of the incorporation of the methyl groups of 'chiral methyl valine' into methylene groups of cephalosporin C, *J.C.S. Chem. Comm.*, 723 (1983).

[133] D. H. G. Crout, M. Lutstorf and P. J. Morgan, *Tetrahedron*, **39,** 3457 (1983).

[134] A. Ozaki, *Isotopic Studies of Heterogeneous Catalysis*, Academic Press, London, 1977.

[135] J. L. Garnett, π-Complex intermediates in homogeneous and heterogeneous catalytic exchange reactions of hydrocarbons and derivatives with metals, *Catalysis Rev.*, **5,** 229 (1971).

[136] H. H. Mantsch, H. Saito and I. C. P. Smith, Deuterium magnetic resonance, applications in chemistry, physics and biology, *Progr. in Nmr Spectroscopy*, **11,** 211 (1977).

[137] K. P. Davis, J. L. Garnett and J. H. O'Keefe, A novel application of electron spin resonance: Selective deuterium exchange using homogeneous and heterogeneous metal catalysts, *J.C.S. Chem. Comm.*, 1672 (1970).

[138] K. P. Davis and J. L. Garnett, A comparison of homogeneous and heterogeneous platinum as catalysts for the deuteration of polyphenyl aromatic hydrocarbons, *Aust. J. Chem.*, **28,** 1713 (1975).

[139] J. L. Garnett and W. A. Sollich, Deuterium exchange reactions with substituted aromatics. 11. The monohalogenated benzenes and naphthalenes, *Aust. J. Chem.*, **14,** 441 (1961).

[140] M. A. Long, J. L. Garnett, R. F. W. Vining and T. Mole, A new simple method for rapid tritium labeling of organics using organoaluminium dihalide catalysts, *J. Am. Chem. Soc.*, **94,** 8632 (1972).

[141] J. L. Garnett, M. A. Long, R. F. W. Vining and T. Mole, A new simple method for rapid, selective aromatic deuteration using organoaluminium dihalide catalysts, *J. Am. Chem. Soc.*, **94,** 5913 (1972).

[142] M. A. Long, J. L. Garnett and R. F. W. Vining, Rapid deuteration and tritiation of organic compounds using organometallic and elemental halides as catalysis, *J.C.S. Perkin II*, 1298 (1975).

[143] M. A. Long, J. L. Garnett and R. F. W. Vining, Tritium labeling of alkanes using organoaluminium dihalide catalysts, *Tetrahedron Lett.*, 4531 (1976).

[144] J. L. Garnett, M. A. Long and A. L. Odell, ^{3}H Nmr and its applications in surface catalysis, *Chem. in Australia*, **47,** 215 (1980).

[145] J. L. Garnett, M. A. Long and C. A. Lukey, Application of ^{3}H n.m.r spectroscopy to metal catalysis orientation of incorporated isotope in halogenated benzenes and alkylbenzenes tritiated by heterogeneous platinum exchange, *J.C.S. Chem. Comm.*, 634 (1979).

[146] J. L. Garnett, M. A. Long, C. A. Lukey and P. G. Williams, Tritium nuclear magnetic resonance studies of the specificity in platinum-catalysed hydrogen isotope exchange of nitrogen heterocyclic compounds, *J.C.S. Perkin II*, 287 (1982).

[147] R. C. Fahey and G. C. Graham, The proton magnetic resonance spectrum of phenanthrene, *J. Phys. Chem.*, **69,** 4417 (1965).

[148] M. A. Long, J. L. Garnet and C. A. Lukey, Tritium labelling of hexamethyldisiloxane for ^{3}H nmr referencing, *Org. Mag. Res.*, **12,** 551 (1979).

[149] M. A. Long, J. L. Garnett, C. A. Lukey and P. G. Williams, Metal-catalysed exchange of alkyl and silyl hydrogen in organosilanes with tritium gas over Raney nickel as measured by ^{3}H n.m.r., *Aust. J. Chem.*, **33,** 1393 (1980).

[150] D. I. Bradshaw, R. B. Noyes and P. B. Wells, Fast catalysis of a hydrogen exchange reaction at low temperatures by gold, *J.C.S. Chem Comm.*, 137 (1975).
[151] D. Calvert, A. Kazekevics and A. L. Odell, to be published (see Ref. 144).
[152] M. A. Long, J. L. Garnett and P. G. Williams, Tritium gas exchange with aromatic and aliphatic hydrocarbons over metal containing ZSM-5 and related zeolites, *Aust. J. Chem.*, **35,** 1057 (1982).
[153] J. A. Osborn, F. H. Jardine, J. F. Young and G. Wilkinson, The preparation and properties of tris(triphenylphosphine)halogeno-rhodium(I) and some reactions thereof including catalytic homogeneous hydrogenation of olefins and acetylenes and their derivatives, *J. Chem. Soc. (A)*., 1711 (1966).
[154] H. B. Kagan and J. C. Fiaud, New approaches in asymmetric synthesis, *Topics in Stereochemistry*, **10,** 175 (1978).
[155] D. Valentine and J. W. Scott, Asymmetric synthesis, *Synthesis*, 329 (1978).
[156] H. B. Kagan and T-P. Dang, Asymmetric catalytic reduction with transition metal complexes. 1. A catalytic system of rhodium(I) with (-)-2,3-*O*-Isopropylidene-2,3-dihydroxy-1,4-bis(diphenylphosphino)butane, a new chiral diphosphine, *J. Amer. Chem. Soc.*, **94,** 6429 (1972).
[157] W. S. Knowles, M. J. Sabacky, B. D. Vineyard and D. J. Weinkauff, Asymmetric hydrogenation with a complex of rhodium and a chiral biphosphine, *J. Am. Chem. Soc.*, **97,** 2567 (1975).
[158] K. Achiwa, Asymmetric hydrogenation with new chiral functionalized biphosphine–rhodium complexes, *J. Am. Chem. Soc.*, **98,** 8265 (1976).
[159] P. M. Hardy, P. W. Sheppard, D. E. Brundish and R. Wade, Tritiated peptides. Part 13. Synthesis of [4,5-^{3}H-Leu^2]-and [3,4-^{3}H-Pro^6]-locust adipokinetic hormone, *J. C. S. Perkin I*, 731 (1983).
[160] G. Baschang, D. E. Brundish, A. Hartmann, J. Stanek and R. Wade, The synthesis of *N*-[^{3}H]acetyl muramyl dipeptides of high specific activity, *J. Label. Compounds and Radiopharmaceuticals*, **20,** 691 (1982).
[161] J. Hine, Carbon dichloride as an intermediate in the basic hydrolysis of chloroform. A mechanism for substitution reactions at a saturated carbon atom, *J. Amer. Chem. Soc.*, **72,** 2438 (1950).
[162] E. J. Forbes and D. C. Warrell, Unpublished results, 1964.
[163] D. S. Kemp, The relative ease of 1,2-proton shifts. The origin of the formyl proton of salicylaldehyde obtained by the Reimer–Tiemann reaction, *J. Am. Chem. Soc.*, **72,** 2438 (1950).
[164] J. G. Atkinson, J. J. Csakvary, G. T. Herbert and R. S. Stuart, Exchange reactions of carboxylic acid salts. A facile preparation of α-deuteriocarboxylic acids, *J. Am. Chem. Soc.*, **90,** 498 (1968).
[165] A. Latif, T. M. Saleh and K. V. Sarnen, *Holzforschung*, **21,** 46 (1967).
[166] G. H. Daub, V. N. Kerr, D. L. Williams and T. W. Whaley, Organic synthesis with stable isotopes, pp. 7–10 in *Proc. Third Intl. Conf. on Stable Isotopes, Argonne National Lab.* (Eds. E. R. Klein and P. D. Klein), Academic Press, New York, 1979.
[167] E. Buncel, A. R. Norris and K. E. Russell, The interaction of aromatic nitro-compounds with bases, *Quart. Rev.*, **22,** 123 (1968).
[168] M. R. Crampton, Meisenheimer Complexes, *Adv. Phys. Org. Chem.*, **7,** 211 (1969).
[169] M. J. Strauss, Anionic Sigma Complexes, *Chem. Revs.*, **70,** 667 (1970).
[170] F. Terrier, Rate and equilibrium studies in Jackson–Meisenheimer complexes, *Chem. Revs.*, **82,** 78 (1982).
[171] E. Buncel, M. R. Crampton, M. J. Strauss and F. Terrier, *Electron Deficient Aromatic- and Heteroaromatic-Base Interactions*, Elsevier, Amsterdam, 1984.
[172] P. Salomaa and K. Sallinen, The stereochemistry of the 2,5-dialkyl derivatives of 1,3-dioxolone(4) and the kinetics of their neutral hydrolysis, *Acta. Chem. Scand.*, **19,** 1054 (1965).
[173] A. Kankaanperä, L. Oinonen and P. Salomaa, Distribution of tritium labeling

between alternative sites in carbon compounds, *Acta. Chem. Scand*, **29A,** 153 (1975).

[174] A. Kankaanperä, L. Oinonen and P. Salomaa, Alternative routes in hydroxide ion-catalysed hydrogen isotope exchange of unsymmetrical ketones, *Acta. Chem. Scand.*, **31A,** 551 (1977).

[175] E. Buncel, A. R. Norris, W. J. Racz and S. E. Taylor, Metal ion-C(8) binding in purine nucleosides. Ready formation of carbon-bound inosine and guanosine complexes of methylmercury(II), *J.C.S. Chem. Comm.*, 562 (1979).

[176] J. L. Garnett and R. J. Hodges, Exchange in aromatic compounds. The labelling of nitrobenzene, bromobenzene, naphthalene and acetophenone, *Chem. Comm.*, 1001 (1967).

[177] D. S. Farrier, J. R. Jones, J. P. Bloxsidge, L. Carroll, J. A. Elvidge and M. Saieed, Catalytic tritiation and ^{3}H nmr spectroscopy of complex organic mixtures—Application to oil shale process waters, *J. Label. Compounds and Radiopharmaceuticals*, **19,** 213 (1982).

[178] D. S. Farrier and J. R. Jones, Concepts and development of a novel radiotracer: U-^{3}H complex organic mixtures, pp. 321–326 in *Proc. Intl. Symp. on Synthesis and Applications of Isotopically Labeled Compounds, Kansas City, 6–11 June 1982* (Eds. W. P. Duncan and A. B. Susan), Elsevier, Amsterdam, 1983.

[179] M. M. Coombs, Carcinogenicity and chemical constitution, *Lab. Technology*, 172 (1983).

[180] P. Sims, P. L. Grover, A. Swaisland, K. Pal and A. Hewer, Metabolic activation of benzo(a)pyrene proceeds by a diolepoxide, *Nature*, **252,** 326 (1974).

[181] S. Lesko, W. Caspary, R. Lorentzen and P. O. P. Ts'o, Enzymic formation of 6-oxobenzo[a]pyrene radical in rat liver homogenates from carcinogenic benzo[a]pyrene, *Biochemistry*, **14,** 3978 (1975).

[182] G. M. Blackburn, A. J. Flavell, P. E. Taussig and J. P. Will, Binding of 7,12-dimethylbenz[a]anthracene to DNA investigated by tritium displacement, *J.C.S. Chem. Comm.*, 358 (1975).

[183] F. Chytil and D. E. Ong, Mediation of retinoic acid-induced growth and anti-tumour activity, *Nature*, **260,** 49 (1976).

[184] R. L. Hale, W. Burger, C. W. Parry and A. A. Liebman, Preparation of high specific activity all *trans*-α-retinyl-11-^{3}H acetate, *J. Label. Compounds and Radiopharmaceuticals*, **13,** 123 (1977).

[185] N. Adachi, J. E. Smith, D. Sklan and DeW. S. Goodman, Radioimmunoassay studies of the tissue distribution and subcellular localization of cellular retinol-binding protein in rats, *J. Biol. Chem.*, **256,** 9471 (1981).

[186] F. Chytil and D. E. Ong, Cellular retinol- and retinoic acid-binding proteins in vitamin A action, *Fed. Proc.*, **38,** 2510 (1979).

[187] E. A. Evans, H. C. Sheppard and J. C. Turner, Validity of tritium tracers. Stability of tritium atoms in purines, pyrimidines, nucleosides and nucleotides, *J. Label. Compounds*, **6,** 76 (1970).

[188] R. Mathur-de Vré, A. J. Bertinchamps and H. J. C. Berendsen, The effects of irradiation on the hydration characteristics of DNA and polynucleotides. 1. An nmr study of frozen H_2O and D_2O solutions, *Radiation Res.*, **68,** 197 (1976).

[189] R. Mathur-de Vré, The nmr studies of water in biological systems, *Progr. Biophys. Mol. Biol.*, **35,** 103 (1979).

[190] R. Mathur-de Vré, R. Grimee-Declerck, P. Lejeune and A. J. Bertinchamps, Hydration of DNA by tritiated water and isotope distribution: A study by ^{1}H, ^{2}H and ^{3}H nmr spectroscopy, *Radiation Res.*, **90,** 441 (1982).

[191] F. Cacace, Gaseous carbonium ions from the decay of tritiated molecules, *Adv. Phys. Org. Chem.*, **8,** 79 (1980).

[192] F. Cacace, β-Decay of tritiated molecules as a tool for studying ion-molecule reactions in interaction between ions and molecules, in *Interactions between Ions and Molecules* (Ed. P. Ausloos), Plenum Press, New York, 1975.

[193] P. Giacomello and M. Speranza, Formation and reactivity of gaseous acetylium ions for the methylation of carbon monoxide, *J. Am. Chem. Soc.*, **99,** 7918 (1977).
[194] P. Giacomello and M. Schueller, Reactions of 'free' methyl cations with aromatic substrates. A study of their attack on *tert*-butylbenzene in the gaseous and liquid phase, *Radiochimica Acta.*, **24,** 111 (1977).
[195] F. Cacace and P. Giacomello, Aromatic substitution in the liquid phase by bona fide free methyl cations. Alkylation of benzene and toluene, *J. Am. Chem. Soc.*, **99,** 5477, (1977).
[196] F. Cacace and P. Giacomello, Aromatic substitutions by [$^3H_3$]methyl decay ions. A comparative study of the gas- and liquid-phase attack on benzene and toluene, *J.C.S. Perkin II*, 652 (1978).
[197] F. Cacace and M. Speranza, Proof of existence of cyclic $C_4H_7{}^+$ ions in the dilute gas phase, *J. Am. Chem. Soc.*, **101,** 1587 (1979).
[198] G. Angelini, M. Spearanza, A. L. Segre and L. J. Altman, Synthesis and ^{3}H nmr analysis of [1,4-$^3H_2$]benzene: A natural source of tritiated phenylium cations, *J. Org. Chem.*, **45,** 3291 (1980).
[199] R. Cipollini and M. Schüller, Preparation, purification, analysis and storage of multitritiated propane, *J. Label. Compounds and Radiopharmaceuticals*, **15,** 703 (1979).
[200] J. W. Emsley and J. C. Lindon, *Nmr Spectroscopy Using Liquid Crystal Solvents*, Pergamon, New York, 1975.
[201] P. Diehl and M. Reinhold, Isotope effects on the degree of order and the deuterium quadrupole coupling constants, as measured by n.m.r. of oriented benzene-d_1, 1,4-benzene-d_2 and 1,3,5-benzene-d_3, *Mol. Phys.*, **36,** 143 (1978).
[202] T. C. Wong and L. J. Altman, Tritium and proton nuclear magnetic resonance study of the isotope effects on the molecular structure and the degree of order of partially oriented [$^3H_1$]benzene, *J. Mag. Res.*, **37,** 285 (1980).
[203] See, for example, J. R. Jones, Solvent isotope effects, Chapter 10 in *The Ionisation of Carbon Acids*, Academic Press, London, 1973.
[204] J. M. A. Al-Rawi, J. P. Bloxsidge, J. A. Elvidge, J. R. Jones and R. A. More O'Ferrall, Applications of tritium nuclear magnetic resonance spectroscopy to the determination of isotopic fractionation factors in methanol–methoxide solutions, *J.C.S. Perkin II*, 1593 (1979).
[205] W. Egan, G. Gunnarsson, T. E. Bull and S. Forsén, A nuclear magnetic resonance study of the intramolecular hydrogen bond in acetylacetone, *J. Am. Chem. Soc.*, **99,** 4568 (1977).
[206] G. Gunnarsson, H. Wennerström, W. Egan and S. Forsén, Proton and deuterium nmr of hydrogen bonds: Relationship between isotope effects and the hydrogen bond potential, *Chem. Phys. Lett.*, **38,** 96 (1976).
[207] L. J. Altman, D. Laungani, G. Gunnarsson, H. Wennerström and S. Forsén, Proton, deuterium and tritium nuclear magnetic resonance of intramolecular hydrogen bonds, isotope effects and the shape of the potential energy function, *J. Am. Chem. Soc.*, **100,** 8264 (1978).
[208] M. L. Huggins, Bond energies and polarities, *J. Am. Chem. Soc.*, **75,** 4123 (1953).

Index of Compounds

Subject Index